FLYAWAY

Also by Suzie Gilbert

HAWK HILL

FLYAWAY

How a *Wild Bird Rehabber*
Sought Adventure and
Found Her Wings

SUZIE GILBERT

With Illustrations by Laura Westlake

An Imprint of HarperCollins*Publishers*
www.harpercollins.com

Grateful acknowledgment is made to reprint the following:

"The Grackle" by Ogden Nash. Copyright © 1942 by Ogden Nash. Reprinted by permission of Curtis Brown, Ltd.

"I Want a New Duck," words and music by Christopher John Hayes, Huey Lewis, and Alfred Matthew Yankovic. Copyright © 1985 WB Music Corp. (ASCAP), Huey Lewis Music (ASCAP), and Lew-Bob-Songs (BMI). All rights on behalf of itself, Huey Lewis Music and Kinda Blue Music. Administered by WB Music Corp. All rights reserved. Used by permission of Alfred Publishing Co., Inc.

"Super Freak," words and music by Rick James and Alonzo Miller. Copyright © 1981 Jobete Music Co., Inc., and Stone Diamond Music Corp. All rights controlled and administered by EMI April Music Inc. and EMI Blackwood Music Inc. All rights reserved. International copyright secured. Used by permission.

FIRST EDITION

Designed by Leah Carlson-Stanisic

Library of Congress Cataloging-in-Publication Data is available upon request.

ISBN: 978-0-06-156312-6

09   10   11   12   13   ID/RRD   10   9   8   7   6   5   4   3   2   1

TO JOHN, MAC, AND SKYE
AND TO ED STOKES

CONTENTS

ILLUSTRATIONS

PART ONE

◦ ◦ ◦ ◦ ◦

Chapter 1

A SECOND CHANCE

The morning sun shone across the Hudson Highlands as I climbed a small wooded hill, dressed in faded jeans and an old shirt and carrying what appeared to be an enormous butterfly net. I carefully scanned the bushes, and within moments found what I was looking for: a large dark bird, one wing hanging haphazardly, huddled next to an old iron fence.

The bird stiffened and eyed me suspiciously. My best chance was to lunge forward and drop the net over his head before he had a chance to run, but before I could do so a small and angry voice cut through the springtime air.

"What are you doing?" it demanded. "What are you going to do to that . . . that animal? Do you have a license?"

I turned to find a diminutive elderly lady standing behind me, her clenched hands on her hips. She was quivering with indignation, and her blue eyes bored into mine.

"I do have a license," I told her, lowering my voice so as not to alarm the bird still further. "I take care of injured wild birds. One of your neighbors called and told me there was one here with a broken wing. I'm going to catch him and take him to the vet and see if we can fix him up."

Undecided, she continued to glare at me; in return, I gave her a genuine smile. I love people like this. Ninety pounds of outrage, she was ready to go to the mat for an injured creature, even though she wasn't exactly sure what it was.

"He's a black vulture," I continued. "He's a cousin to those big turkey vultures, the ones you always see circling above town. Vultures are great birds—I'm happy to take care of him while he recuperates."

She stared at me doubtfully, making up her mind. "Well," she said finally, "just don't hurt him. Remember—I'll be watching you."

As it turned out she wasn't the only one; by this time four neighbors had gathered behind us. I started toward the vulture, hoping he was tired and hungry and would stay crouched in the leaves so I could net him quickly and efficiently. But according to Murphy's Law of Wildlife Rescue, this only happens when no one is around to admire your skill. Whenever there's a crowd, whatever bird you're after will spring to life and lead you on a chase designed to make you look like an incompetent fool. Naturally, that was what happened here.

Most people don't think of vultures as being particularly nimble; but in reality they can run like jackrabbits. Trailing his broken wing he sped around the fence, only to come face to face with an impenetrable tangle of barberry. "Excuse me!" I called to the four neighbors. "If he comes toward you, block his path!"

I lunged toward my quarry and brought the net down, but the vulture was no longer there. Having feinted right, he ducked left and raced toward the neighbors, who scattered like confetti in the wind. Triumphantly, the vulture slipped by and hightailed it down the road. I let out a whine of dismay and glanced at the elderly lady, who scowled at me disapprovingly. Clutching my net, I ran after the bird's retreating form.

I spent the next half hour running through what seemed like every backyard in the small river town. Had the circumstances been different I might have enjoyed seeing its variety: the small 1950s houses and the stately restored Victorians, the perfectly tended gardens and the areas of cheerful chaos. I kept careening around corners, gasping for air, just in time to see the disappearing edge of a black tail feather. At one point the vulture tore through an alleyway and I took a swipe at him with the net; beak open, one long black wing fully extended, he leaped upward and landed on a stairway railing just as the lady of the house opened the door. Letting out an ear-shattering scream, she threw herself back inside; something breakable crashed to the floor, and I staggered on.

The end of the line came in a surprisingly large, almost empty backyard. By this time the two of us must have looked like the fox and the hound in the old cartoon, where the chase continues even though both are so exhausted they're walking instead of running. As the vulture made a final sprint across the lawn I dashed after him, extended my net, tripped, and fell forward through the air. Hitting the ground with a thud, I looked up to see the bird safely, miraculously, enclosed in my net. We both lay on the grass, our sides heaving, listening to the crows screaming above us.

Finally a man's voice made me look up. "Excuse me?" it said. "Do you need some help?"

I gazed at him for a moment, sorely tempted to say something ungrateful. "Thank you," I said instead. "I have a red Jeep parked down on Violet Street. There is a pile of towels in the back—could you bring me one?"

"Sure thing," he said, and jogged off.

The air was fragrant, the late spring sunshine warm. I sat up and regarded the vulture encased in my net. If the fracture was in the middle of the humerus, the large wing bone closest to the body, odds are it would heal well and he would eventually be released. A fracture close to the joint would be more difficult, the prognosis unclear. A fracture involving the joint usually means the bird will never fly again. But whatever the outcome, at least now he had a second chance.

I felt a creeping sense of well-being. I wasn't a conquering hero, but I had saved this bird from a sure death by either starvation or predation. I would return to my car, mission accomplished, and perhaps the assembled crowd—if they were still there—would feel a new appreciation for the wildlife around them. My helpful friend returned and, smiling, handed me a towel. I extricated the big, dark bird from the net and held him briefly, allowing the man to see the obvious bond between the avian world and me. The vulture looked me in the eye, opened his beak, and with a master's timing, regurgitated the contents of his stomach onto my lap. I looked up; the man was no longer smiling.

Vomiting as a defense mechanism makes perfect sense for a vulture. Vultures are nature's clean-up crew, and their insides are a marvel of engineering: they can actually eat a victim of hog cholera and not get sick. What goes down the hatch isn't normally all that appealing, and when marinated in those formidable gastric juices and hurled back out again it's even less so. The product doesn't actually have to touch you, either—just landing in your general vicinity is enough to make most normal creatures respect the vulture's wish to be left alone. There hadn't been all that much in this vulture's stomach, but what there was was especially aromatic.

"Jeeeeeesus!" the man burst out. "He sure loves *you*, don't he?"

I wrapped the bird in the towel like a large papoose and carried him back to the car, where a small crowd waited. As I approached their noses began to wrinkle; looks were exchanged. Standing to one side was the elderly lady, arms crossed. When I stopped in front of her she squinted at the vulture, her face twitching slightly; she took an almost imperceptible step backward.

"I got him," I said. "I'll take him to the vet and she'll set his wing, then I'll take care of him until he recovers. If he can be released I'll bring him right back here and let him go, okay?"

"Well," said the lady with a small smile. "Maybe not *right* back here."

Chapter 2

THE SERPENTINE ROAD

My mother fed the wild birds. She filled our feeders with seed and hung blocks of suet from the maple tree outside the kitchen window, hoping to coax the cardinals and the woodpeckers into view. When I was twelve I combed through the Nature section of the local library and discovered the British book *Hand-Taming Wild Birds at the Feeder*. Inspired, I began setting my alarm for 6 A.M. As my father left for work he would find me standing motionless beneath the maple tree, seed-filled hand outstretched, waiting for a stouthearted chickadee or titmouse to dart across my palm.

I found that if I relaxed to the point where I was barely breathing the birds became more confident, and in time they would perch on my finger and even stare into my face before snatching a seed and flying away. They were nearly weightless, and unlike my dogs and horses, gerbils and rabbits, geckos and guinea pigs, there was nothing that bound them to me. They could abandon me at any moment, a realization that left me both disturbed and exhilarated. Domestic animals were familiar and dependable; wild birds were their opposite: mysterious and otherworldly, and my ability to call them down from the trees was an unexpected and thrilling feat of sorcery.

Unfortunately, my magical powers did not extend to convincing the chickadees to accompany me to boarding school, where I routinely snuck away from the suffocating company of other teenaged girls and hitched a ride to a nearby stable owned by a former jockey. Gray-haired, chain-smoking, crippled by a long-ago fall, Sam would wait until everyone else had gone home, then pull out two glasses, a bottle of Seagram's and a can of 7-Up, and mesmerize me with fast-paced tales of honest horses and crooked bets. Filled with whiskey and camaraderie, I'd climb onto one of his rangy jumpers bareback while he limped out to the main ring and raised the fences. It was foolish and dangerous, but faced with a wall of gaily colored rails, the horses rose into the air like birds in flight—as if they were in on the game, as if they had no intention of being part of a limp, a scar, or a career left in pieces.

Eventually I had to leave them all behind. I bounced from college to college, sometimes leaving by my own free will and sometimes not. I traveled from country to country and job to job, educated and socially adept, but far from where I could call the birds down from the trees or count on the kindness of horses. By my mid-twenties I'd decided to buckle down and get a real job in New York City, until it became apparent—time after time—that I was incapable of sustained office work. There was my typing, which was slow; my dictation, which was nonexistent; my attitude, which I was told was poor; and my "problem with authority figures," which, first noted by a boarding-school headmaster and repeated by two college deans, was then echoed by a long succession of bosses.

Finally abandoning the idea of an office job I started walking large packs of dogs, a successful niche halted by my first marriage. Faced with a new life of shopping and decorating I retreated to the northern end of Central Park, the dumping ground for unwanted dogs, where I would catch a glimpse of the gaunt and shadowy creatures before they disappeared into the woods. If I sat still for long enough one would eventually make its way toward me, hostile and haunted, used to abuse but desperate for food, willing to risk human contact for a pocketful of kibble. Sooner or later it would accompany me home, and the apartment slowly filled with stray dogs, all but one en route to permanent homes.

Not surprisingly the marriage failed, and an old college friend invited my one remaining dog and me to live with her on her farm in Maine. I'll come for the weekend, I said, and I stayed nearly two years, tucking myself in, safe among the farm animals and finally free to walk the empty, overgrown fields. After the din and turmoil of the city the small farmhouse was a refuge, entering the quiet hay-filled barn like meditation. In winter the snow rose outside the windows and slowed the pace of the day, enveloping the house in a pale and protective silence.

The spring after I arrived my dog, a big husky/shepherd mix, heard the call of the wild. I spotted him through the kitchen window, a white wolf with blood-soaked jaws, running down a sheep while three others lay motionless nearby. I loved the sheep even though I knew they would eventually end up in someone's freezer; I loved my dog even though he had committed what was, at first glance, a terrible act. That night I agonized over his aggression, yet I remembered when he had savaged a man who tried to attack me in Central Park and I called him my hero and bought him a steak dinner. A New York City official would have condemned him to death for attacking a person; a Maine farmer would have praised him for defending a family member, but shot him for killing the sheep.

It all reinforced my belief that humans were perverse and their decisions arbitrary, that animals were always true to themselves but at the mercy of a species not qualified to be in charge. I replaced the sheep, chained my dog whenever the farm animals were in the field, and made sure that his reputation as a livestock killer went no farther than our front door.

I might still be in Maine if it weren't for John Horgan, a New York–based science writer with an adventurous spirit and a dicey résumé, who managed against all odds to entice me back to the city. But when I returned, New York seemed even larger and more oppressive than when I left. My beloved killer dog died at the relatively advanced age of thirteen, sending me into a storm of mourning that John tried to assuage with the gift of a young yellow-collared macaw. It was a thoughtful and effective gift, but it opened yet another can of worms. We moved to a small town in the Hudson Valley, where I landed a job in an animal hospital, and before long started bringing home abused and un-wanted parrots. The small rented house echoed with jungle shrieks and John started to crack.

"You're killing me," he said. "I'm going deaf and they're chewing the house to pieces. Just find homes for these last few and then wait until we move into our own place, okay?"

We bought a house sitting alone on the edge of a 1,200-acre rarely used state park. On our first visit I glanced around the interior and then went back out the door, walked down the first hill of the long dirt driveway, crossed a wide barberry-choked field, and plunged into the woods, which were filled with towering trees and rocky outcroppings and streams that wound their way down to the Hudson River. According to my map, the woods went on for miles. There were no neighbors. It was perfect.

Soon a friend who knew of my interest in birds sent me to the Hudson Valley Raptor Center, at that point a rehabilitation facility for injured birds of prey. I hovered outside a huge outdoor flight cage filled with red-tailed hawks, awed by the regal, lethal creatures who seemed unbowed by their encounters with humankind. But in the clinic was an owl who had been poisoned, an eagle who had been shot, a vulture who had been hit by a car—each crouched and subdued as it struggled to survive. I felt a familiar rush of sympathy mixed with anger. Despite the two-hour round trip, I volunteered at least once a week for the next eleven years, during which time I somehow ended up having two children.

My friends and family were shocked. My only explanation was that, like the falcons and the vultures and the harriers, I eventually bowed to the bio-logical imperative. Children had always filled me with alarm, a reaction that escalated after the birth of my own. Far from friends and family, alone with two young children and clueless as to how to raise them, I lived in fear that they would somehow discover that all I wanted to do was flee. But we muddled through, mostly thanks to the twice-weekly daycare that allowed me to return to the raptor center. My chore of choice was maintaining the outdoor enclosures; equipped with rubber gloves, a rake, scrub brush, buckets of water, and a wheelbarrow filled with defrosted rats, I'd spend hours alone with the birds who had entranced me, my stress and fatigue draining away.

Ninety-five percent of the injuries suffered by wildlife are the direct result of human activity; the trouble with working at a wildlife rehabilitation center is that you get to see the unending parade of damage firsthand. After several years I decided to try to help prevent the breakage, not just pick up the pieces. Fresh from creating the center's newsletter, I contacted several local newspapers to see if they would be interested in publishing an environmental column written by Elizabeth T. Vulture, an ornery, unreleasable turkey vulture who actually resided at the raptor center. I supplied a head shot of Elizabeth, several columns, and a list of potential topics. A small chain of upstate papers took the bait, and Elizabeth—snide, sarcastic, prone to black humor, and unimpressed with the human race—had a monthly column.

Elizabeth ranted and raved about pesticides, poisons, outdoor cats, habitat destruction, electrocution of raptors on utility poles, pigeon shoots, predator control, real estate developers, right-wing congressmen, and Wise Use movement members; during lighter moments she included feather-wearing in women's fashion, those who use dissected animals to create "art," and PMS (pre-migrating syndrome). The real Elizabeth became a local celebrity, turning her back on those who tried to photograph her and snaking her bald head through the bars of her flight cage in an attempt to bite her fans. Things went swimmingly until the owner, a good liberal environmentalist, sold the newspaper chain. Soon after, I received a call from the editor.

"Listen," he said. "Elizabeth is going to have to . . . uh . . . tone it down."

"What?" I said. "Why?"

"It's the new owners," he said. "Now we have a legal department. They said they're afraid that if they print that last piece you wrote, Monsanto will sue them."

"You're not serious," I said.

"I am serious," he replied.

"They're *afraid* Monsanto will sue them?" I burst out. "They should *hope* Monsanto will sue them! A mother of two who lives in the woods and takes care of hurt birdies and writes as a vulture for a tiny chain of newspapers gets sued by the huge evil chemical conglomerate that brought us Agent Orange and Frankenfoods and DDT? The company that nearly caused the extinction of our national bird? Are you kidding me?"

"I know, I know," said the editor. "If it were up to me . . ."

"What do they think they are, the *New York Times*?" I railed. "I'll tell you what—you convince them to print it, then I'll send it to Monsanto myself with a cover letter saying 'Go ahead and sue me, you bastards! I can't wait to see a head shot of your CEO next to a head shot of Elizabeth! Who do you think will win this one?' "

As it turned out, Monsanto won. The new owners insisted that the company would sue them, not me, and said that if I wanted to continue to write for them I had to be "nicer." There went the writing gig.

I contacted a New York agent, who read a stack of clippings and said she could get me a weekly column. It was a hard decision. I might be able to foam at the mouth entertainingly, but I preferred to sit alone in a flight cage filled with birds of prey. I wanted to spread the environmental word, but I was afraid a weekly column would take me away from my children—both of whom, I was still convinced, were doomed by having me as a mother. I put the column on hold for my family, whose personalities seemed to grow more extreme by the day.

There was John, who wrote deliberately controversial science articles, then chortled happily over his resulting hate mail, and who explained being fired from *Scientific American* magazine after the publication of his book *The End of Science* by saying, "I guess the marketing department didn't think it was funny."

There was our son Mac, who by age six had an almost mystical connection with birds and liked to sit cross-legged at the local Buddhist monastery, yet who was obsessed with machine guns and ceremonial swords. There was our daughter Skye, seventeen months younger, who would careen through her days at the speed of sound, ripping cabinet doors off their hinges, launching herself from the tops of bookshelves, and during the occasional meditative moment, absently chewing on electrical cords. There was Zack, the swaggering little yellow-collared macaw, who would bite our guests, then laugh uproariously in a voice suspiciously like mine. And there was Mario, the recently rescued African grey parrot, who would soar through the house whistling old Motown songs, searching for wallets and important letters he could chew to pieces. Encircled by this daily maelstrom, I was even more grateful for my quiet and solitary moments with the wild birds.

But all nonprofit organizations, especially those run almost exclusively by volunteers, are subject to fluctuation and change, and often philosophical differences can be solved only by a parting of ways. After I left the organization I went through raptor withdrawal, staring longingly at the occasional hawk soaring over my head, knowing the nearest alternative raptor center was over an hour and a half away. The solution came in a phone call from a local lady who had heard that I worked with birds, who called to ask if I could help two swans imbedded with fishhooks and trailing fishing line. Later that day I sat on my deck—scratched, bruised, exhausted, covered with marsh muck, and speckled with ticks—and thought: this is great. I'll rescue and take care of injured wild birds at home. I'll set up a small, local one-person operation and my bird world will be steady, self-contained, and completely under my control. All the decisions will be mine, and mine alone.

It seemed like a plan.

Chapter 3

PROBLEMS AND SOLUTIONS

Now that I am older, wiser, and more haggard, I look back on my decision to rehabilitate wild birds at home with incredulity. There is only one sane way to get your wild animal fix: by volunteering at a bird or wildlife center. You show up, you work hard, you go home, you resume your life. Your wildlife work may occasionally spill over into your regular life, but it will not engulf it like a tidal wave, which is what happens when you attempt to set up shop at home.

This fact of life had been explained to me by several veteran rehabilitators, all of whom burst into gales of laughter when I said I was going to combine bird rehab with family life.

"How old are your kids?" said one, wiping her eyes.

"I'm going to start next spring," I said firmly. "They'll be seven and eight."

"Hmmm," she said, assuming a perplexed expression. "Was I going to feed the kids and worm the crows—or vice versa?"

"It'll be great for your marriage," added another. "Men just love women covered in bird doo."

"I'm already covered in bird doo," I said. "I have parrots."

A third bugged her eyes and stared maniacally into space. " 'I'm ready for my close-up, Mr. DeMille,' " she hissed. " 'Just get these owls off me.' "

I scoffed at their lurid predictions. People create their own destiny, I had always thought; you could weasel your way into or out of any situation, given the right motivation. The key was to be specific. Where there was a problem, there was a solution.

I swung into action.

The problem: how to start a wild bird rehabilitation operation, at home, from scratch. I contacted the bird rehabilitators in my area, of whom there were surprisingly few, and found that what was desperately needed—besides more rehabilitators—was a good flight cage. Flight cages are the large outdoor enclosures where recovering adults can regain their wing strength and juveniles can learn to fly before they are released. At that point I was familiar only with raptor flights, the enormous enclosures made of evenly spaced wooden slats. Federal regulations prescribe flight-cage size according to each species; a flight cage for a red-tailed hawk, the most common raptor species in my area, must be at least ten feet by fifty feet by twelve feet.

It didn't take me long to figure out that building a proper raptor flight cage was a pipe dream. Our property's terrain is hilly, rocky, and heavily wooded; a few phone calls revealed that I couldn't even afford to build the flight, let alone clear the trees and bulldoze the hills so I had somewhere to put it. But at least an injured raptor could go to the raptor center; from what I had heard, there were no flight cages available for the injured waterbirds and songbirds of my area. I thought fondly of the swans I had just freed from their web of lines and fishhooks. The problem was that recuperating waterbirds eventually need water, and our pond was located at the very edge of the property, far from the house and right next to the road.

The solution: I would build a flight cage for songbirds only. Since I couldn't accommodate the birds I had come to know, I would return to the birds of my childhood—the small perching birds of gardens and backyards. I had never lost the feeling I had experienced when I first stood outside my house, seed-filled hand outstretched and a chickadee hovering inches away. I still viewed even captive wild birds as mysterious and otherworldly, essentially untamable, my brief proximity to them a rare and fragile gift. If songbirds were the neediest birds in my area, then a songbird flight cage was what I would build.

The problem: I needed a clinic. Early on I actually considered breaking through the wall of our bedroom and adding a small bird hospital room, complete with heat and running water. A rehabber friend, to whom I will be forever grateful, seized me by the shoulders and said firmly, "Just get a gun and shoot yourself in the head. It would be quicker."

The hospital room also fell victim to financial reality. As I searched for alternatives, I regarded myself critically. I had a husband, two young children, and two parrots; common sense dictated that there were only so many creatures I could care for at once. But I knew what I was prone to and, worse, what I was capable of. I needed parameters set in stone, not subject to the vagaries of chance and my own bad influence.

The solution: I would not take injured birds at all. I would build a songbird flight cage and announce that I would take in only small birds from other rehabbers—adult birds who had recovered from their injuries and just needed conditioning, or juveniles who simply needed to practice flying before release. Once I had my license and the Department of Environmental Conservation asked if they could give my name out to the public, I would say no. By removing the clinic, I actually believed that I was removing the one thing that would allow my bird operation to spiral out of control.

The problem: how to learn to care for songbirds, who have neither talons nor any desire to eat defrosted rats. I bought books. I borrowed books. I surfed the net and printed out information. I joined Wildlife Rehab, an electronic mailing list that encompasses all wildlife, but I set up my account so I received only e-mail regarding wild birds. Electronic mailing lists are a godsend for rehabbers, especially single ones working out of their homes. Once you join, you are linked with rehabilitators from all over the country—sometimes from all over the world— and whenever a member of the group posts an e-mail, you receive it. Subscribers include newcomers and veterans, single rehabbers and those working in wildlife centers and zoos, specialists who deal with only one type of bird and those who deal with whatever comes through the door. For example, someone posts a question, "What is the best diet and setup for a hooded warbler with a broken leg?"

and inevitably another writes back, "I've done *hundreds* of hooded warblers! My *middle name* is Hooded Warbler!" and showers the subscribers with advice and tips, which I would dutifully print out and file alphabetically in a purple three-ring binder labeled "SPECIES SPECIFIC."

The solution: sweat equity, the currency of rehabilitators everywhere. I spent months helping a friend who rehabs all kinds of birds, including the ridiculously small ones. During one of my first visits she showed me how to hold an injured chipping sparrow (weight, 10 grams).

"Look here," she said. "You see that thing on his foot?"

"His foot!" I said. "I can barely see the bird."

The problem: how to build a songbird flight cage when the only flights I had ever seen were for raptors. I went on field trips. I visited several bird rehabilitation centers, took photos, photocopied their blueprints, and interviewed the volunteers about what they would change if they could. Songbird flight cages are smaller than those built for raptors, but the entire enclosure must be encased in metal hardware cloth (which is like chicken wire but stronger and has small squares) and lined with soft mesh. I asked questions: A-frame versus straight rectangle? Loft or no loft? What was the best substrate?

The solution: It was an A-frame with a small loft, encased in half-inch hardware cloth, lined with plastic mesh, and had natural flooring with added organic soil and wood chips.

The problem: where to put the flight cage. It needed to be near the house, but not too near the house. It needed sun, but not too much sun. Wherever it was built would entail cutting down some trees, but I hoped not too many trees. It couldn't be built on rock ledge—which probably lay beneath half our property—because a trench a foot and a half deep would have to be dug around the perimeter of the cage, the hardware cloth rolled downward and angled out and weighted with rocks, all to deter digging predators.

The solution: 150 feet southwest of the house, tucked into a small valley between two hills. If the flight cage were angled properly, it would be protected from the north wind and receive dappled sunlight throughout the day. There were rocks, but they were removable. The area was large enough so I could expand the size of the flight—maybe even build two. And the only tree that would need to come down was a huge old dying oak that had been struck by lightning and was already listing alarmingly to one side.

The possibility of two flight cages and the avoidance of healthy tree slaughter: this was becoming intoxicating.

The problem: who would build the songbird flight cage? (It wasn't going to be me.) I grabbed the telephone. Had I been more plugged in to the local home-building scene, I would never have had the nerve to call Bruce Donohue and Michael Chandler, who, unbeknownst to me, were renowned for their high-end, elegant craftsmanship. As it turned out, they were also faithful environmentalists with a soft spot for nonprofit work. They were enthusiastic about aiding the recovery of injured wild birds, and they happened to have a small hole in their schedule.

The solution: Bruce and Michael looked at the site, studied the blueprints, gave me a generous break on their fee, and said they could have it up in a week. I was the proverbial snowball rolling down the hill.

The problem: I needed both a New York State Wildlife Rehabilitator's license and a federal permit to rehabilitate migratory birds. A state license allows the rehabilitator to care for mammals, reptiles, amphibians, and nonnative birds (house sparrows, starlings, and pigeons). Potential rehabbers have to pass a 100-question, multiple-choice test covering the natural history of all local species of wildlife, as well as their emergency care, nutrition, restraint techniques, wound management, parasitic infections, epizootic and zoonotic diseases, and release criteria.

A federal permit allows a person to rehabilitate all native birds. Obtaining the permit entails writing an autobiographical summary of your avian expertise; describing what your birds will be fed and how you will obtain specialized foods; submitting diagrams and photographs of your facilities; and gathering letters of recommendation from what seems like every person on the planet who has ever uttered the word *bird*.

The solution: For the state test, I studied. In most areas, wild opossums live two to three years. The only sure way to kill the eggs of the raccoon roundworm is with a blowtorch. Feed kitten milk replacer to orphaned bobcats and goat's milk to orphaned white-tailed deer. For the federal permit, I wrote. I called. I asked people to say nice things about me. I sent out stamped, self-addressed envelopes. I rolled my eyes. I said bad words.

◎ ◎ ◎ ◎ ◎

In mid-September, the kids and I sat on our deck listening to the clatter of hammers against wood. As the flight cage rose in the distance the kids casually tossed me state license questions, proving once again that young brains absorb information far more quickly than older ones.

"What do you call the underside of a turtle's shell?" asked Mac.

"The carapace," I replied.

"Wrong!" crowed Mac. "It's the plastron!"

"Darn!" I said. "Well, at least I know that a rabbit isn't a rodent."

"But that one's easy," said Skye, sighing deeply. "Everyone knows that rabbits are lagomorphs."

The flight cage was more solid than my own house. It was a 400-square-foot enclosure separated by a plywood wall into two rectangles twenty feet long, ten feet wide, and eight feet high, covered by an A-frame roof and lined with ethereal green mesh. In the late afternoon sun it looked magical, a place where a broken bird could learn to fly again, a temporary refuge created by a crew who were craftsmen by day and artists and musicians at night. It had just come into the world, but already its karma was good.

Where there was a problem, there was a solution. I was confident that when our doors opened in the spring, everything would go according to plan.

Chapter 4

GENESIS

chaos theory (ka'os' the'e-re): a theory that complex natural systems obey rules but are so sensitive that small initial changes can cause unexpected final results, thus giving an impression of randomness. (MSN *Encarta*)

There are many areas to which chaos theory applies. In wildlife rehabilitation, it rules.

For the single, home-based rehabilitator in particular, one's sensitive complex system hinges on the ability to set limits. As both logic and my longtime rehabber friends told me, one has a finite amount of space, a finite amount of energy, and a finite number of hours in the day. Small changes in one's limits can cause unexpected results, especially when they become the norm instead of the exception.

Eventually I would come up with my own version of chaos theory.

chaos theory (ka'os' the'e-re) (2): a theory that the more confidence a bird rehabilitator has in her ability to manage her facility, the more rapidly it will slide into chaos. **chaos** (ka'os'): a condition or place of great disorder or confusion. (*The American Heritage Dictionary*)

In the beginning, however, I was golden.

◎ ◎ ◎ ◎ ◎

It was early May and we were all in the kitchen, waiting for the delivery of our first songbird. The kitchen is a spacious room with a sliding glass door that opens onto the deck. Ninety degrees from the glass door is a bay window, on the bottom of which rest two large manzanita parrot jungle gyms. The jungle gyms are three feet tall and covered with toys and ropes, with food and water dishes bolted to the highest perches. The one on the left belongs to Zack, the yellow-collared macaw, and the one on the right to Mario, the African grey parrot.

Long ago I tried to make the window more "tropical" by placing several nontoxic potted plants around the bases of the jungle gyms. The parrots froze and gazed at me in horror, as if they had just watched me seed their territory with nuclear warheads and could not get over the betrayal. Since this was a somewhat normal reaction I left the room for an hour, figuring they'd get over it. I returned to find that they had rappelled downward, yanked the plants out of the pots, chewed them to pieces, flung the dirt all over the bay window, and then pushed the pots onto the floor, where they lay broken into what appeared to be several million pieces. The window bottom is now covered with miscellaneous unbreakable parrot toys, but little else.

"When is she going to get here?" demanded Skye, hopping up and down. At seven years old, she was kinetically active, emotional, and theatrical, prone to racing through the house singing Christmas carols at the top of her lungs, no matter what the season.

"She'll be here," said Mac, who at eight was calm, centered, and sported a mane of blond hair that reached past his shoulders. Having started growing it in homage to his medieval, dragon-riding heroes, he had stubbornly refused to cut it despite the taunts of his schoolmates, none of whom had hair past their ears.

"What kind of bird is it, again?" asked John, leaning against the counter in rip-riddled running gear. After the successful publication of his first book, John had left his job intending to set up a home office and write another, only to realize that there wasn't enough room in the house for him to do so. The eventual solution was a small writing cabin tucked between the house and the flight cage, affordable as long as a bathroom was not included. With the construction of his cabin and my flight cage, it seemed as if we were both on the road to our dreams: living in a beautiful rural area, each doing what we cared about, connected to the advantages of modern life but raising our family as far from the stress and hustle of it as we could manage.

"It's a house finch," I said. "He dislocated his wing a few weeks ago, it was set and healed, and now he just needs to start flying again."

House finches are small brownish-gray flocking birds, often seen around backyard feeders; the males have reddish-orange feathers from their heads down to their bellies. This finch was being delivered by Maggie Ciarcia, whom

I had met at a workshop for prospective wildlife rehabilitators. Sponsored by an organized network of rehabbers in the county south of where I live, the workshop had starred Maggie and Joanne Dreeben, both veterans, who spent two hours trying to ensnare innocent civilians. They showed slides, told stories, referred to graphs, and introduced their wildlife ambassador: a huge, easygoing white pigeon. No matter where you go in the world there are never enough wildlife rehabilitators, which is not surprising: there is little or no money in it, and like all nonprofit work, the burnout rate is high. But every once in a while someone will hear the call of the wild, and another rehabber is born.

"I take small mammals, small birds, and game birds," Maggie had said, smiling, to her rapt audience. "Lots of baby bunnies and squirrels. Songbirds and pigeons. I love wild turkeys—they're my favorites. No raptors. I don't take any birds that can hurt me. Joanne will, though; she'll take anything."

The unflappable Joanne shrugged and gave a rueful grin. "Whatever," she said.

Both had been delighted with the prospect of a new flight cage, and they had exchanged knowing glances when I told them firmly that I was not taking any nestlings or injured birds—only healthy adults ready for a flight cage.

"Yeah, good luck with that," said Joanne.

There was the barely audible sound of a car door slamming, and both kids bolted from the room. We heard the front door open, Maggie's voice greeting the kids, and a moment later she was escorted into the kitchen. Zack let out a piercing shriek, bowed energetically, flashed his eyes, and shouted, "Hello!" Mario, more circumspect, regarded the new human suspiciously and withheld comment.

"Did you bring him? Is he in there?" demanded Skye, vibrating with excitement and trying to peer into the cardboard carrier Maggie was holding.

"I don't think she'd drive all the way over here and bring us an empty box," said Mac.

"Not to worry," said Maggie. "I guarantee you that whenever I show up here with a box there will be a bird in it."

"And so it begins," said John. "Will this be the best of times or the worst of times?"

Maggie grinned at him. "Probably both," she said.

We all trooped outside, past John's office and to the finch's temporary new home. John and the kids took up positions outside while Maggie and I entered the flight cage. I had furnished both sides with hanging tree limbs and leafy branches, and I planted a live tree in each. On the ground were logs, small brush piles, and shallow rubber saucers for drinking and bathing. In each flight one end of a long, slender tree limb rested on the ground and the other on a perch, allowing birds who couldn't fly to hop up to a comfortable spot. The connecting door between the two sections swung easily and latched securely. There must be something I've forgotten, I thought to myself as I opened the finch's carrier.

The diminutive bird looked up, jumped onto the rim of the carrier, then hopped down onto the ground. He paused briefly to assess his new surroundings, looking like a tiny ringmaster dwarfed by a cavernous Big Top. Making no attempt to fly, he scooted up the angled limb and onto a leafy branch, where he regarded us with concern.

"Is that it?" asked Skye incredulously from outside. "Is that all there is?"

"That's all for now," I said, as Maggie and I filed out of the flight cage and latched the door.

Later that night I turned to John. "I know this is unprofessional of me," I said. "But I hope he's all right out there. He's all alone in a new place. I hope he's not scared. Do you think he's lonely?"

John smiled, blissfully unaware of how many times he and I would have similar exchanges over the next five years.

"I'm sure he's fine," he said. "He's a wild bird."

Chapter 5

THE ONE EXCEPTION

After four days in the flight cage, the house finch was eating well and seemed comfortable, but had made no attempt to fly. I was wary of chasing him, envisioning my very first songbird launching himself off the branch, crashing to the ground, and breaking his formerly dislocated wing. With this in mind I called and made an appointment with Dr. Alan Peterson, a good friend who had promised to donate his veterinary services should I ever need his help.

"A house finch?" asked the receptionist. "You mean a wild bird? I don't think Dr. Peterson sees wild birds. Hold on, let me check."

After a minute she returned. "When would you like to come in?" she asked.

There is no one more important to a rehabilitator's professional life than a skilled veterinarian, and those willing to help injured wildlife are few and far between. Treating wildlife requires an added layer of knowledge, as treatments that may work for domestic animals don't necessarily work for wild ones. Veterinary medicine, often cited as being far more difficult than human medicine because the patient can't say where it hurts, becomes even harder when the patient has no owner to report its recent behavior. And, of course, no owner to pay the bills.

I once heard a story about a young veterinarian, enthusiastic and altruistic but inexperienced with wildlife, who had agreed to treat an injured screech owl that a Good Samaritan had dropped off at his office. The vet carried the cardboard box into one of his exam rooms, peered inside, and was dismayed to find that the owl had died. He picked up the owl, a tiny creature about five inches tall, and laid it gently on his table. Unaware that screech owls play dead when stressed, he leaned over to give it a closer look, whereupon the owl sprang to life, launched itself off the table, and sank its talons into the vet's nose. Unable to remove the owl himself he bellowed for his technicians, but they were at the other end of the hospital and couldn't hear him. He ended up having to walk through the crowded waiting room with a small owl hanging from his nose, which reportedly he did with surprising aplomb. I don't know the eventual outcome of the story; I would hope that this experience cemented his relationship with wildlife, although it could very well have had the opposite effect.

Later that day I stood in one of Alan's exam rooms, watching as he opened the cardboard carrier a half inch, peered downward, and slowly reached inside. When he removed his hand he was expertly cradling the finch. Alan gazed blandly at the bird, then at me.

"Kernels of corn," he said. "You're rehabbing kernels of corn."

"Really!" I said, returning his look. "So are you."

Alan can't help himself. It's his nature to look for the absurdity in every situation, and as soon as he finds it, he feels compelled to point it out. Unfortunately, as soon as he points it out I feel compelled to start arguing with him, even if I secretly believe that his view may be valid. Such was the case here.

The person who had found the finch had gone out of his way to make sure the little bird arrived safely at a veterinarian's office. The veterinarian had donated his expensive time to examine and treat the bird, then had called Maggie. Maggie had driven to the office, picked up the bird, then spent three and a half weeks feeding and caring for him. She had delivered him to me and now here I was, consulting a second normally well-paid veterinarian, whose advice would certainly include an indeterminate number of weeks of additional food and care. Should everything go well, I would eventually drive the bird back to his original location for his release.

And it was all for a house finch, a common, 23-gram species of songbird. In terms of time, money, and effort—not to mention gasoline—it might have seemed a bit absurd.

Except that it wasn't.

Heaving a weary sigh, Alan went to work. He gently felt along the finch's wing, almost imperceptibly moving each joint. "Stiff," he said. "He needs a little physical therapy. Move it like this—back and forth—very slowly. When you feel any resistance, stop. Twice a day for a few days. Then try tossing him into the air. Gently. Just make sure he has something to land on besides the ground."

"Great!" I said. "Thanks, Alan!"

"No problem," said Alan. "That'll be eight hundred dollars."

The following day I caught the little finch and slowly manipulated his wing, struck anew by the fragility of songbirds and their ability to survive despite the obstacles that humans so carelessly throw into their paths. He was a trouper, wearing a resigned expression as I slowly moved his wing back and forth. And he did look resigned, despite the fact that birds' faces are fixed and supposedly cannot show expression. The problem is not with the average bird's inability to show expression; the problem is with the average human's inability to perceive subtlety. Perhaps if birds had giant eyebrows to waggle and fleshy lips to distort, they'd be easier to figure out.

After a few days I gathered armloads of brush from a nearby field, piled it under one of the hanging branches, and then tossed the finch into the air, hoping that he wouldn't need to use his makeshift landing pad. He flew halfway across the flight and landed gracefully on the branch, sending me into paroxysms of glee. A few days later he wasn't flying as well, sending me into the depths of gloom. I can't keep this up, I thought, suddenly appreciating another advantage to working at a wildlife center: instant emotional support. If a grounded bird starts flying, you celebrate with your compatriots; if he takes a turn for the worse, you share the pain. In my case John was gone for the day, so I had to wait for my support team to return home from elementary school.

"That little finch wasn't flying so well today," I said, after they'd jumped off the school bus and were accompanying me up our long dirt driveway.

"What?" Skye gasped, looking stricken. "Is he going to die?"

"No!" I said quickly. "He's only. . . ."

"How do you know?" she demanded. "How do you know he's not lying dead on the ground right this minute?"

"He's not dead," said Mac. "And he's not going to die, either. He's probably just tired from all that flying."

"There you go," I said, adding the final link to our emotional daisy chain. "He's going to be fine."

Soon all I needed to do was to walk toward the finch and raise my hand, and he'd launch himself from his branch and fly to another. I kept wishing for another finch to keep him company—not that I wanted another bird to be injured, but if one *were* to be injured I wished that he would find his way here. This was where I made a serious "wish error." As anyone who has ever heard a fairy tale can attest, all wish genies get a big kick out of messing with wishers who are not specific. I wished for another finch, and suddenly one appeared. But it was not a house finch.

It was an American goldfinch.

What the heck! one might reason. Goldfinches are in the same family as house finches. At least it's a finch. That was my reaction.

At first.

I was unprepared for the phone call. A friend of a friend had found the goldfinch, dazed and motionless, outside one of her windows. She had placed it carefully into a cardboard box; when she opened the box a half hour later, it hadn't moved. Could you please take him? she asked. He's so beautiful and I don't know what to do for him. I can drive him right over.

I hesitated, and a war broke out inside my head. I had spent a year working out the master plan, each subplan, and every individual detail. I had set my rehabilitation rules in stone: no injured birds. No birds from anyone but other rehabbers. No birds that couldn't go right into the flight cage. I envisioned my rules as bowling pins and the goldfinch as a speeding ball, heading down the center line. I can't crack this early in the game, I thought.

"Uhhhhh," I said. "Actually, I'm not really set up for . . . I'm in a . . . kind of . . . ."

It is almost impossible to predict how an impact injury, whether it be from a window or a car, will turn out. Minutes after the injury some birds are raring to go; others are dead. Some have broken bones or spinal injuries along with their head trauma, others appear to have no ill effects. Some will be bruised and in pain but appear to be improving, then three days later they'll die when the blood clot you can't see reaches a certain part of their brain. Many rehabbers keep birds who have suffered head trauma for a period of time after the injury even if they seem to be fine, just as an added safety measure.

I had no local number to give the woman with the goldfinch. Maggie and Joanne were the only bird rehabilitators within an hour of me, and both were at work. I did have a contingency plan for emergencies: I could keep one or two injured songbirds temporarily in pet carriers—which I had collected over the years and stored in a small stacked tower in the garage—in our extra bathroom. I could take the goldfinch, give him a small dose of the anti-inflammatory Maggie had given me, and if he wasn't better in a couple of hours I could take him to Maggie. If he recovered quickly, he could go into the flight cage with the house finch.

"Is driving the bird an hour and a half away an option?" I asked. "Because if not, maybe I could . . . ."

I made a silent vow: it would be only this once.

"Never mind," I said. "Just bring him over."

The goldfinch looked like a bird of the tropics, a dazzling combination of deep black and brilliant yellow. I gave him his medicine, put him into a small pet carrier, and left him alone in the extra bathroom for an hour. When I returned he had hopped off the floor of the towel-covered carrier and was sitting up on a perch.

"Hey," said John at dinner that night. "What's that bird doing in the bath-room?"

"There's a bird in the bathroom?" asked Skye. "Can I see him?"

"Is he hurt?" asked Mac. "What kind of a bird is it?"

"It's a goldfinch," I said. "He hit a window, but he's better now. He has to rest right now, but you can see him tomorrow morning before you go to school."

"I thought you weren't taking injured birds," said John.

"I'm not," I said firmly.

⊙ ⊙ ⊙ ⊙ ⊙

The following morning the goldfinch was active and alert, and had eaten all his niger seed and most of his mixed seed, so I released him into the flight cage. The house finch brightened visibly at the sight of another bird, rustling his feathers and watching closely as the goldfinch surveyed his new surroundings. After ten minutes I left them alone and walked back to the house, just in time to see Maggie getting out of her Jeep.

"Did you get my message?" she asked, reaching into her car and pulling out a small cardboard carrier. "One more for your flight cage. I thought I could drop him off on my way to work—can you help me take off his bandage?"

"No problem," I said. "Step into my office."

We went into the extra bathroom, where it would be easy to retrieve the bird should he escape while being handled. As I closed the door Maggie pulled a leather glove out of her jacket pocket, put it on, then started to open the cardboard carrier.

"What have you got in there?" I asked, puzzled.

"A house sparrow," she replied.

"A *house sparrow*?" I said, grinning. "You want me to get you a whip and a chair?"

"Observe," she said, holding her gloved hand up, index finger pointing toward the ceiling. Opening the carrier with her other hand, she ceremoniously dipped her index finger into its depths, then slowly removed it. Clamped onto the end was a small brown bird with a bandaged wing, its stout beak determinedly grinding away at the glove, its feet pedaling furiously through the air. Maggie reached into her pocket and pulled out a cotton washcloth, and in one smooth motion enveloped the sparrow in the washcloth and pulled it away from her glove.

"I'd watch this thing if I were you," she said. "He'll bite your finger off."

As Maggie cushioned the sparrow inside the washcloth I flipped up one corner and removed the bandage. "Uh-oh," I said, grimacing at the slightly swollen and hardened wing. "It doesn't look too good."

"Ouch!" said Maggie, who had just been pinched through the washcloth. "I can take him back to the vet who set it, but I can't get there until the weekend."

"Tell you what," I said. "Leave him here and I'll see if I can run him up to

Alan tomorrow. The only thing is . . . it's a house sparrow. Alan will kill me for bringing him a house sparrow."

"Maybe we could get some magic markers and disguise it as a goldfinch," said Maggie.

"Here's where things get tricky," I said.

Chapter 6

QUANDARIES

For thousands of years groups of humans have arrived in places they don't belong, muscled their way past the native population, taken over the territory, then flooded the area with their own hordes of descendants. Humans who do this are called "settlers." Occasionally they bring along birds who don't belong, either. When the birds follow the lead of the humans, they are called "invasive species."

The male house sparrow is a stocky, strikingly patterned bird with a black bib and a thick, formidable beak. His more somberly dressed mate is a member of the LBJs (little brown jobs), the large group of basic brown sparrows between which new birders despair of ever being able to differentiate. Brought over from England somewhere around 1850, house sparrows quickly spread across the United States, preferring areas of human habitation. Both male and female are surprisingly aggressive, especially when it comes to taking over other birds' nests; they will actually kill the nestlings of gentle native species such as bluebirds and swallows—and sometimes the parents as well—earning them the enmity of birders everywhere.

"My anglophilia probably does not include English sparrows," e-mailed my friend Ed, "but they, too, must individually appreciate some TLC."

For someone who is as electronically inept as I am I've done surprisingly well by the Internet, which has supplied me with libraries of information and connected me with many wonderful and helpful people, including one who became a dear friend: Ed Stokes, a nature lover, birder, sailboat designer, and sled-dog enthusiast who, at one point, lived with eighteen Siberian huskies. Ed is also a writer and philosopher who can take any complicated situation and boil it down to one simple, elegant sentence, as he did with my house sparrow conundrum.

"Thanks, Ed," I wrote back. "It's not that I want to encourage the rotten little bluebird killers, it's just that this one happens to be in my flight cage and has a bad wing."

At that point it was easy for me to be cavalier. I had not had the experience of building, setting up, and monitoring a field of bluebird houses, hoping to help a species whose numbers are rapidly dwindling, only to check them one morning and find occupant after occupant slaughtered by house sparrows, whose numbers are rapidly increasing. At that point, however, I didn't see a moral quandary; I saw only a single bird in need of care.

The following afternoon I entered the flight cage with a small towel and a cardboard pet carrier. The house finch hopped across several branches and flew to a high limb, while the goldfinch motored back and forth with surprising speed. I turned away from both of them and set my sights on the flightless sparrow, who was on the ground pecking at seeds. Tossing a dark towel over a bird normally causes it to crouch down in the sudden darkness, giving you time to bend over and pick it up. I tossed the towel over the sparrow, and a nanosecond later he sped out from under it and shot over to the other side of the cage.

I left the flight cage and walked over to John's writing cabin, where he was working on his second book. Filled with file cabinets, bookshelves, old record albums, and various eclectic and eccentric memorabilia, the cabin looks like a miniature Adirondack house nestled between two hills of mountain laurel.

"I hate to disturb you," I said, sticking my head in the door. "But could you help me for a minute? There is no way I can catch this bird by myself."

Soon we were both in the flight cage, flinging towels and brandishing long-handled nets while the sparrow avoided us with the energy and velocity of a pinball. Finally we held the towels like matadors and, dragging them on the ground, herded him into a corner; faced with the inevitable he hurled himself into the air, came down headfirst in the dirt, and digging furiously, attempted to tunnel his way away from us.

"You have to hand it to them," said John, shaking his head. "That's why they're taking over the world."

◎ ◎ ◎ ◎ ◎

Alan flashed me a grin, closed the door behind him, and looked down at his file.

"So!" he said. "You described our last patient as an 'adult male house finch with a formerly dislocated left wing, still unable to fly.' Today's patient is . . . let's see . . . how did you put it? Here it is: 'injured bird.' Hmm. A strangely vague description."

"Vague but accurate," I said helpfully.

"Care to elaborate?" he asked.

"Compound fracture," I replied. "Originally brought to another rehabber. Wrapped for two weeks. I took the bandage off yesterday—wing is hard and swollen. Feels like a solid mass."

"Okay," he said, putting his hand inside the carrier. "And what sort of bird are we dealing with?"

It was the moment of truth. It was high-stakes poker. I looked him in the eye.

"A little one," I said.

Alan withdrew his hand and regarded the bird.

"A house sparrow," he declared. "And it just bit me."

Alan and his wife, veterinarian Jan Robinson, are both avid birders. They win birdathons, during which teams of birders travel from site to site trying to identify the most species within a twenty-four-hour period. They bird by ear, sometimes needing only two notes to identify a small warbler otherwise invisible to mere mortals. They are active in their local bird club, of which Alan is a former president. They work to preserve habitat for native species.

They put up bluebird houses.

"The bone is infected," said Alan, after giving the sparrow a thorough exam. "The only chance he has is if I amputate the wing. I don't know how far it's spread, so even if I take the wing off he could still die of the infection.

"It's up to you," he finished, giving me a level gaze. "I told you I'd help you and I will. Just tell me what you want me to do."

Every rehabilitator has his or her own way of dealing with life-or-death decisions. There are hard-liners on both sides: those who feel that it is always kinder to euthanize a wild bird than to take away its freedom, and those who feel that they must try to save every bird no matter how terrible its injury or how miserable its life in captivity will be. The majority fall somewhere in the middle, always trying to assess the situation and the individual, always trying to keep the emotions at bay and do what is best for the bird.

Finding a home for an unreleasable bird is not easy. A captive bird needs space, light, companionship of its own kind, and freedom from pain and fear. Some birds never adjust to captivity and live out their lives as prisoners who have committed no crime. Some birds' injuries require constant attention, which can sometimes be in short supply at a busy wildlife center.

Removing the entire wing from a wild bird is now illegal, although it wasn't when I was trying to decide what to do with the house sparrow. The reasons it is now illegal were valid back then, though: an amputation that close to the body doesn't heal well, and the bird will always be off balance. The sparrow's surgery would be difficult, his recovery painful, and should he survive, his long-term chances were questionable. Recovery would mean isolation, a distressing state for a flocking bird. And if he did recover, he would need a good permanent home with other house sparrows.

On the other hand, he was a feisty little bird and I would be saddened by his loss.

In this case, I was lucky. I didn't have to spend hours agonizing over what to do, weighing alternatives, making a list of pros and cons only to discover that they came out even; it was obvious that putting a wild bird through such a grueling ordeal for a chancy outcome was unfair, that I would be doing it more for myself than for the bird.

I sighed. "Can you put him to sleep?" I asked.

"Sure," said Alan, picking the sparrow up once again and cradling him gently. He paused. "I'm sorry," he said.

"Thanks," I said.

I drove home, my throat tight. Common, invasive, aggressive, he was still a miraculous creature once capable of flight, a tiny bird I knew only briefly but who left an impression so indelible that years later I can instantly call him to mind.

I remembered a long-ago county fair where I had listened to a wildlife rehabilitator talk about his work. "Wild birds hide their troubles," the soft-spoken man had said. "By the time we get them, most of them are heading for the edge. What we do is simple: we bring them back, then we let them go."

He had paused, waiting for the audience's smiling reaction. How easy! How satisfying! Then he delivered the punch line. "The only problems are the exceptions," he said. "And the exceptions will occur in nearly every case you get."

Bring them back, then let them go. But there are different ways of letting go.

When I reached my driveway I stopped and waited for the school bus, debating what to tell the kids. How rosy a picture should you paint for a seven- and an eight-year-old? What circumstances warrant a lie? Maybe, I decided, I will just avoid the subject.

"How's the sparrow?" they both asked as they piled into the car.

"I'm sorry," I said, and told them the truth, carefully explaining the reasons behind my difficult decision. Mac looked crestfallen; Skye looked at me uncomprehendingly, her eyes welling with tears.

"But Mommy," she said. "You're supposed to help them, not kill them."

It took me a moment to respond. "I know, honey," I said finally. "But sometimes that's the only way I can help them."

We drove home in silence.

Chapter 7

THE OTHER EXCEPTIONS

"Suzie!" bellowed a familiar voice. "Where are you? Come out here! I have a bird for you!"

When Ruth summons you, you respond, especially if she is shouting at you from your own living room.

"There I was, walking Rolfie through that field down by the river," she explained, gesturing animatedly out the window toward the large long-haired German shepherd circling the house. "And I see these two robins, and one is kicking the crap out of the other one! Really! Beating the bejesus out of him! So I figure I can't leave him there with his wing hanging down and all, so I picked him up and put him in the back of the car! And he's out there right now waiting for you!"

Years before, my writer/editor friend Robert Hutchinson and I were part of a small group locked in battle with a developer who wanted to build an $80 million retirement complex in the middle of our tiny hamlet. A typical e-mail from Robert would include a prospective strategy based on a section of Sun Tzu's *The Art of War*, a proverb in the original German, several outrageous puns, an aside on the geological anomalies located beneath the land in question, the coda of an ancient Celtic battle hymn, and a scathing critique of the developer's haircut. Soon after the developer was sent packing I invited Robert and his wife, whom I had never met, to dinner. Five minutes before they arrived I had an anxiety attack.

"Why didn't I invite other people as a cushion?" I said to John. "Opposites attract. She's going to be some mousy little thing that never opens her mouth, and we're all going to end up sitting there staring at each other."

Robert's wife Ruth turned out to be the former Ruth Copeland, a blues singer from a small town in northern England who arrived in this country in the early sixties, immediately embarking on an odyssey that included dodging police bullets during the Detroit race riots and touring the country with Parliament-Funkadelic and Sly & the Family Stone. Ruth launched into a description of a concert where she belted out rock anthems while wearing a buckskin bikini and full Indian headdress and swinging from a trapeze above a packed house; Robert countered with a fond remembrance of chasing a neighborhood drug dealer, who had punched Ruth over a pay-phone dispute, down Broadway with a meat hammer.

"Come on, then," said Ruth, "come and get this bird out of my car."

I hesitated. I had made one exception, and was not about to make another.

"I'm going to give you my friend Maggie's number, and you call her and she'll take the robin."

"What?" said Ruth, staring at me as if I'd sprouted a second head. "Why would I call someone else? You're the bird lady, and I've brought you a bird."

"But Ruth," I said firmly. "I'm not set up for injured birds, I'm only taking the ones that are healthy enough to go out into the flight cage."

"Right," she said, rolling her eyes and pushing me toward the door. "Well. Once you fix his wing I'm sure he'll be healthy enough to go into the flight cage. Now hurry up and get him out of my bloody car before he covers it with guano."

One more exception, I told myself. Only this one. And only because the bird was from a personal friend who wouldn't take no for an answer.

I could feel the fracture in the robin's humerus. But I wasn't sure if that was the only damage, and driving the ninety-minute round trip to Alan's office was simply not in the cards on this Saturday morning. However, the vet who treated Maggie's injured wildlife divided her time between her own mobile veterinary service and the Cortlandt Animal Hospital, which is a relatively short distance from my house. It was worth a try.

"She said to bring the robin in," replied Janet Hartmann, the hospital receptionist who had listened sympathetically to my problem. "How's twelve-thirty?"

Wendy Westrom, VMD, knows her way around wildlife, having started out as a staff veterinarian at the Bronx Zoo. There she met her future husband, Dr. Fred Koontz, who was then the curator of mammals. Fred and Wendy's rented apartment became the unofficial rehab area for a long and varied list of ailing creatures, among them an abandoned baby sea lion that the resourceful and unflappable Wendy smuggled into an unused hot tub downstairs. Wendy is incredibly generous with her skills, manages to see the bright side of any situation, and excels at calming anyone distraught over wildlife, all of which have earned her the undying gratitude of countless rehabbers and members of the public.

Wendy X-rayed and wrapped the robin's wing, chatted with me about bird rehab, told me she was at the hospital on Wednesdays, Fridays, and some Saturdays, and offered to help me with any birds that needed veterinary care. "Maggie tells me you have two beautiful flight cages," she said.

I nodded. "I really only take birds that are ready to fly. I have two kids, and I don't have the space for injured birds."

Wendy gestured to the robin. "You mean except for that one?" she asked with a grin.

"Yeah," I said. "Except for that one."

"I have one that a rehabber just dropped off," said Wendy, "He's unreleasable and going to a sanctuary on Long Island. They're going to pick him up in a couple of days. Could he stay in one of your flights? It would be a lot nicer for him than my basement."

"Sure!" I said.

As Wendy left the room it occurred to me that I hadn't asked her what kind of bird it was. Did "I have one" mean another robin or another bird? Wendy returned and placed a very large cardboard box on the floor. From within the box came an audible rustling, then a loud thud. Wendy saw my expression.

"It's a redtail," she said. "Is that okay?"

I hesitated. I had already made two exceptions, and was not about to make another. But Wendy had just done me a favor and was so generous about offering her help. The red-tailed hawk was staying only a couple of days and would require little care.

And I love redtails.

"It's okay," I said.

⊙   ⊙   ⊙   ⊙   ⊙

Red-tailed hawks are a relatively common species, found throughout the United States and readily seen soaring above a field or surveying the road from a tall tree. Their call is the famous hoarse descending screech that Hollywood always equates with "bird" no matter where the movie takes place or what kind of bird actually appears on the screen. Redtails are big and fierce but matter-of-fact; once they settle into a rehab situation they can be mellow and easy patients. And this particular redtail wasn't even a patient: his wing, broken near the shoulder joint, had already healed, leaving him unable to fly but otherwise perfectly healthy.

I carried the box to the flight cage, congratulating myself on having the foresight to ask Bruce and Michael's crew to put up one large raptor perch—just in case—before they finished construction. A thick dowel covered with outdoor carpeting and bolted at an angle to one corner, it needed only a ramp to provide the flightless redtail with access to a comfortable perch. I hunted around until I found a thick, solid tree limb that had fallen to the ground, and dragging it into the flight cage, I leaned it against one side of the perch and secured it with rope. I took down two of the hanging branches, not wanting the hawk to be tempted to try to go for them.

The two finches were in the second flight. I opened the dividing door, stuck my head in to check on them, and nearly lost my eye to the speeding goldfinch, who hurtled past me in a brilliant yellow blur. After pulling the door shut I opened the cardboard box and tipped it over slowly, allowing the redtail to trot out by himself. He looked around, climbed up the ramp, and settled onto the perch.

I headed back to the car for the robin but made a quick detour to the freezer in the garage. Bought to store the spillover from our small kitchen unit, it was filled with the usual assortment of foods except for the bottom drawer, which held two bags of large rats. They were a gift from a friend at the local zoo, who had received a big shipment of raptor food and knew I was starting a bird rehab operation.

"You say you're only doing songbirds," she had said, "but I know how that goes."

I pulled one of the rats out, carried it into the kitchen, locked it into a new freezer bag, and slid it into a plastic tub filled with hot water. I put the whole thing on top of the washing machine, which was hidden behind a set of folding doors, and covered it carefully with a dishtowel. Returning to the car, I removed the robin's carrier and brought it into the extra bathroom.

"Mommy!" shouted Skye from upstairs. "Is that you? Where is my blue sweatshirt?"

"It's me," I called back. "And it's folded on the dryer."

I heard her thundering down the stairs, loudly caroling her way through the month of May. "Deck the halls with boughs of holly, fa la la la la, la la la la . . ."

I stopped. No, there was no way she'd notice. It was covered with a dishtowel.

"Tis the season to be . . ."

The song was interrupted by an ear-splitting scream. I bolted from the bathroom.

"What's the matter?" shouted John, rushing down the stairs with Mac on his heels.

Skye stood rigid, eyes wide, staring at the top of the washing machine.

"There's a giant mouse up there!" she gasped. "And he's dead in a freezer bag!"

John and Mac followed her gaze and grimaced. Then, as one, all three turned and looked at me.

"Uhhh," I said. "It's, um, actually, it's a rat. It's food. For the redtail."

"Redtail!" said Mac. "Where'd you get a redtail?"

"Well!" I replied, "I brought the robin to the vet, and as it turned out . . ."

"Wait a minute!" John interrupted. "Where'd you get the rat?"

"From the freezer in the garage," I said.

John looked aghast. "There are rats in our freezer?" he said.

I decided to take the offensive. "They've been there for two months!" I shot back. "So I think it's a little late to start getting upset!"

"But what about the mouse?" wailed Skye.

Wildlife rehabilitation is a moral minefield, beginning with the food you serve the recuperating patients. Raptors, as well as many other birds, need to eat whole animals in order to stay healthy. You can buy mice and rats, but the cost is prohibitive. Although I would venture to say that while very few rehabilitators support animal testing, most who care for raptors receive mice and rats from companies that breed them to sell to laboratories. The companies donate the extra or imperfect ones, already dead, to rehabilitators. You can rail about animal testing and feel badly about the death of so many animals; thinking that the dead are the lucky ones doesn't help. But the bottom line is that without these donations, nursing certain birds back to health would be impossible.

It was something I had wrestled and come to terms with years ago. It was something I would eventually explain to my kids. But as we were all sitting around the kitchen table, under the watchful eyes of the parrots, I felt the subject was something I could put it off a bit longer.

"Listen," I said. "I know I said I was only taking songbirds, but sometimes things get a little complicated and I end up with, say, a redtail, and the redtail needs to eat a rat. So I get the rats that are already frozen from the zoo, and I just put them in the freezer and take them out when I need them. And they're all in bags and they're in a separate drawer in the freezer so you don't have to worry about cooties or anything."

"But where did the rats come from?" said Skye.

"From the zoo," I answered, knowing fully well what she meant.

"But where did the zoo get them?" she pressed.

"I'm not sure," I said evasively. "I'll have to find out."

"Probably they were already dead when the zoo got them," said Mac. "And if they're dead, they're dead, and that's pretty much all there is to it."

"Do me a favor," said John. "Defrost them in the garage, all right?"

"The poor rat," said Skye. "I think we should have a funeral and bury him."

"We *are* going to bury him," said Mac. "Inside the redtail."

Zack let out a jungle shriek.

"I couldn't have said it better," said John.

Chapter 8

COOPERATION,
AND LACK THEREOF

I opened the red three-ring binder emblazoned "INCOMING" and turned to the second entry, where I had written "Adult Goldfinch" as the headline. Beneath it was the date I had accepted the bird and a description of its injury. Below that was the name, address, and telephone number of the finder, and in subsequent lines, the weight of the bird, the treatment it received, and a few notes on its behavior and recovery.

At the end of the year all rehabilitators must file a report with the state. Those who rehab migratory birds must file reports with both their state and the federal government. This entails filling out copious forms listing all the above information, as well as the bird's eventual fate. The previous year's tallies are due at the end of January, and the mountains of paperwork required are enough to make the most mild-mannered rehabber swear like a dockworker.

But the logic behind them is sound. Ward Stone, the New York State Wildlife Pathologist, used records of poisoned birds compiled by rehabilitators to help build his case against the pesticide diazinon, whose use was subsequently banned in New York State. Rehabilitators' records can be used to help track outbreaks of diseases such as botulism and West Nile virus. On the unfortunate side, not all rehabilitators have the birds' best interests in mind; governmental overview is one attempt to keep rehabbers from warehousing crippled birds they don't have the "heart" to euthanize.

Keeping notes on each bird is an invaluable way to compile your own database, something to which you can refer to when faced with a similar injury or species. I try to make my notes as detailed as possible, but occasionally brevity has its place. That morning, when I opened my binder for the telephone number of the woman who had found the goldfinch, I glanced at the previous night's entry: "Bird is nuts. Release tomorrow ASAP."

"Normally I would keep him a little longer," I told the woman on the phone. "But he's flying around like a maniac and I'm afraid he's going to hurt himself. I'll bring him back to your house in about a half hour."

"Thank you so much!" she exclaimed. "I'm so happy he's okay."

I entered the first flight cage with a net and a carrier, and I paused to check on the redtail. He had eaten most of his meal and was on his perch, watching me with pale yellow eyes. He was a beautiful hawk, with off-white feathers cascading down his chest and with a dark belly-band. His tail was still brown and streaked, as he was too young to have gone through the molt necessary to attain the red tail feathers distinctive of adults. He was big and healthy, had made it through his first winter, and should have been soaring through the spring sky. But because he had been hit by a car, and because his wing was fractured too badly to heal perfectly, he would spend the rest of his life in a cage.

He was young and would probably adjust. The sanctuary taking him had nice facilities and a dedicated staff. But he was another bird who would never fly again, who would see the sky only through the bars of a cage. The sanctuaries and rehabilitation centers that house healthy but unreleasable wildlife provide many of the public with their first views of a wild creature, and it is an invaluable service. The sight of a hawk up close can change minds and opinions and can win converts to the environmental side. I myself had worked with unreleasable education birds who seemed to revel in their new life, who obviously enjoyed working in front of a crowd and playing up to their audience.

But I still cannot see a flightless wild bird, even one who has spent years in captivity and seems perfectly content, without feeling a pang of sadness. Had I stopped long enough to consider all the complexities I was bringing with me before I immersed myself in wild bird rehabilitation, I might have been able to forecast my emotional future. But it wouldn't have mattered, anyway, because what I wanted the most back then was to be near wild birds, any wild birds, and to try to help them find their way home.

I opened the door to the second flight cage, set the net and the carrier on the ground, and pulled the door closed behind me. As it turned out, I need not have worried about the goldfinch hurting himself; what I should have worried about was his hurting the house finch, which was sporting a fresh wound below his right eye.

Perhaps the house finch, whose flight was still erratic, flew into a branch and wounded himself. But I turned and addressed the goldfinch, who was circling the flight cage like an angry hornet.

"You!" I said. "You're out of here, you ingrate!"

It was easier said than done. I caught the wounded finch and brought him into the house, where I cleaned and medicated the area under his eye and set him up in a carrier next to the robin. I returned to the flight cage and tried to catch the goldfinch; not only was I unable to catch him, I couldn't get within six feet of him. When John came to my aid the goldfinch expertly threaded his way between us.

"We have to stop," I said finally. "He's going to die of stress."

"And then he won't be releasable!" added John.

Although it is normally preferable to return birds to their home territory, there are times when it is not feasible. This was one of them. The alternative— to open the door and simply let the bird go—suddenly highlighted a flaw in my perfect flight cage: the second flight had no outside door. I transferred the redtail to a large crate, then left both the dividing door and the first flight's outside door open, figuring the goldfinch would simply release himself. No dice. Flying energetic laps around his enclosure, he steadfastly refused to go through the dividing door. I returned to the house, confident that when I returned he would be gone.

"There were two phone calls for you," said Mac when I walked in.

One was from a volunteer from the sanctuary on Long Island, who said he would pick up the redtail the following morning. The second was from a rehabilitator named Jenifer Bowman.

"I'm a veterinary assistant at Somers Animal Hospital, and a friend of Maggie's," said Jen when I called her back. "A woman just brought us a grackle she raised. She fed him bugs, so he's pretty healthy. But she raised him alone, so he's imprinted."

When it comes to raising baby birds, the most important thing is a proper diet, followed by the company of the bird's own kind. A baby bird raised alone will grow up relating to humans and not to other birds. A single bird raised by humans is like a single human raised by chimpanzees: while it may survive, its social skills will be sadly lacking. And just as most chimps can't show a human how to run to the supermarket for a bag of groceries, most humans can't show a bird how to forage for food in the wild.

"I have a friend at the zoo who has a grackle almost the same age," Jen continued. "Would you have room to put them together? I was thinking maybe they'd get along and sooner or later you could soft-release them."

A *soft release* is when you let the bird go but continue to provide it with daily food while it acclimates to the outside world—as opposed to a *hard release*, when the bird is completely on his own after he flies off. Soft-releasing an imprinted passerine (any non-raptor perching bird) is a contentious issue. One camp says no imprinted passerine should ever be released; they don't have the skills necessary for life in the wild and are too acclimated to humans to adapt. The other camp says certain survival instincts are hardwired, and under the right circumstances—if the young birds have spent time with their own kind and have had the proper training—some of them can be released.

"I can give it a try," I said. "Just give me a couple of days, though—my flights are a little backed up right now."

By the time I returned to the flight cage three hours had passed. That goldfinch is probably in Arizona by now, I thought, as I started to pull the dividing door shut so I could let the redtail out of his crate. But before I could shut it completely, a bright flash of motion sped by.

He was still there. And, impossible though it seemed, he appeared to be moving even faster than before. Minutes later I was balancing on a ladder, cutting through the metal hardware cloth several inches below the roof of the second flight cage with a pair of wire cutters. I cut three sides of a rectangle, then peeled the hardware cloth downward, creating a roomy opening. Pulling a large pair of scissors out of my back pocket, I sliced an identical opening through the green mesh lining. I peered inside; for once, the goldfinch was still. But not for long.

I started to climb down the ladder, and before I'd even reached the ground, a streak of yellow shot out over my head. And instantly, it was gone.

I stood blinking. It was my very first songbird release. What happened? I had planned to be surrounded by enthusiastic, admiring family members whose lives would be enriched by the experience, one of whom would be taking a sequence of perfect photos suitable for framing. I had planned to release a bird who, once let go, would leap into the air, circle back, dip his wings at me in gratitude, and then soar away into the clear blue sky.

Instead I was the sole participant in the release of an ornery, ungrateful creature who really didn't need my help, who had been a terrible pain in the neck, and who parted without a backward glance.

I had to laugh at the lovely image I had created, an image I knew from the start had nothing to do with reality. The goldfinch had simply followed one of the principal rules of wildlife rehab: they never do what you want them to do.

◦   ◦   ◦   ◦   ◦

Late that afternoon I sat at the kitchen table, overseeing the kids' art projects while on the phone I described the flight-cage surgery to my friend Matt Mc-

Mahon, a bird lover who had offered his carpentry skills for any small projects that came along.

"That's no problem," said Matt. "What I'll do is frame out the rectangle and rest it on a couple of hinges, then put a lock at the top. It'll be like a trapdoor—when you want to release a bird you'll open the lock and pull it down, and when you're done you just push it back closed and lock it again. Tomorrow morning okay? I'll swing by early and get the measurements."

I happily busied myself with paper and glue until Mac broke the silence.

"Look!" he said. "What is that bird doing?"

We followed Mac's pointing finger to the bird feeder filled with niger seed, where the usual mixed group of wild house finches and goldfinches had collected. Normally it was a fairly peaceable kingdom, with the occasional skirmishes between males resolved after a few moments of sparring. But this time a brilliant yellow male appeared to be harassing the others. Flying in furious loops he dove in and out among them, sending them scurrying away while he ricocheted from tree to tree and back again.

"I'm sorry!" I called to the other birds. "I'm so, so sorry."

Chapter 9

PREDATORS, GUARDIANS, AND GRUBS

I needed to take a run.

I'm not a particularly fast runner, nor do I cover dozens of miles at a stretch. I run because I need to get out into the woods, to be alone on a rocky trail, to hear the sounds of the forest without the ubiquitous background hum of traffic and human voices. I also run because if I am stressed out and don't run, I'm afraid my head will explode.

When we first moved into our house we heard stories about a pair of reclusive but ferocious hawks who nested near one of the trails. They were seen only during the late spring and early summer, but woe betide the unwary hiker or runner who finds him or herself on the wrong trail at the wrong time. A neighbor who lived a quarter mile or so down the road from us had achieved local celebrity status by having his head bloodied several times, events he would recount with a grin and a good-natured shrug.

As it turned out, the birds were northern goshawks.

Goshawks, along with Cooper's and sharp-shinned hawks, are members of the accipiter family, a group of long-tailed, short-winged forest dwellers who prey mostly on other birds. Coops and sharpies can be the bane of a songbird lover's existence; no dopes, they will occasionally set up shop around a busy bird feeder and simply pick off the songbirds when they appear. Goshawks, on the other hand, avoid developed areas, preferring uninhabited dense woods. Evidently the goshawks figure they'll stay away from people's houses as long as people stay away from their woods. But people inevitably fail to live up to this bargain, and the trouble begins.

I knew approximately where their nest was located, so when I went for my run that morning I decided to try to find it. I spotted the large nest, the product of years of additions and renovations, wedged in firmly near the top of a dying hemlock tree forty yards or so from the trail. Woolly adelgids, aphid-like insects unwittingly imported from Japan and first discovered in the United States in 1985, have all but destroyed the hemlocks in much of the East, turning dense green forests into areas so blighted they look like the scene of a forest fire. The goshawks' territory was still filled with healthy oak, maple, walnut, and birch, but the area surrounding their nest was mostly dead and dying hemlock. Although the trees that once gave them a thick curtain of privacy were now crumbling around them, the goshawks were unwilling to abandon the nest they had used for so many years.

The nest was there, but its occupants were not. I scanned the trees, found nothing, and started to resume my run. I had taken only a few steps when I heard a ringing cry, ascending in pitch and momentum, that silenced the other sounds of the forest: *kek-kek-kek-kek-kek-kek!*

It was a sound so wild, so stirring, yet so viscerally fear-inspiring that I stopped in my tracks. Those who believe that humans have no trace of wilderness left in their veins should listen to the cry of an angry goshawk, which can instantly reduce the smug descendants of thousands of years of civilization to small, trembling prey. I searched but could not find the source, and I didn't hear the cry again. Finally, knowing I was being watched, I headed for home.

While most people's protective instincts are aroused by cuddly creatures such as puppies and ducklings, mine are also triggered by homicidal raptors with records of assault. Although I wanted to go back to the nesting area the next day, I knew that the female would be sitting on eggs and shouldn't be disturbed. I marked out a month on my calendar; I would return when the eggs had hatched.

In the extra bathroom, the house finch's wound was healing, and the robin who had lost the bird fight was eating like a prize hog and biting me whenever he had the opportunity. The young redtail had gone off to the Long Island sanctuary, and my first worm delivery had finally arrived.

Most birds eat bugs, so bird rehabilitators must always have bugs on hand. Mealworms come in small, medium, large, and super, the latter being so big that they can actually bite you through your pant leg. Some birds prefer wax-worms, which are small, white, soft-bodied grubs that resemble maggots. You can buy a small plastic container of worms at the local pet supermarket for an astronomical amount of money, or you can order them by the thousands from companies such as Grubco or Nature's Bounty, which charge reasonable rates and will ship them to your door.

I couldn't resist a company with a name like Grubco, so I ordered 1,000 medium mealworms and 500 waxworms, which arrived in two cardboard boxes riddled with dime-size airholes. Inside the first box was a muslin bag tied shut with a wire twist-tie. Inside the bag were five or six sheets of crumpled newspaper, and within the newspaper were the mealworms. When I held up the bag I heard the sound of soft rustling, like gentle rain.

In the second box were four light blue plastic containers, each poked with airholes and filled with soft wood shavings, and each containing 125 wax-worms.

"Kids!" I called. "Come on! The worms are here!"

Mac hurried down the stairs. As he looked at me expectantly, the large lump beneath his shirt started moving slowly across his chest.

"Ouch!" he said. "Get out of there, Zack!"

He pulled the collar away from his neck and a small head appeared, beady eyes flashing. The yellow-collared macaw climbed out of Mac's shirt and onto his shoulder, eyeing me and laughing giddily, as if he wanted nothing more than to have me join in the fun. I was wise to this little strategy, however; it meant that if I put my hands anywhere near Mac, Zack would rush down like a velociraptor and try to bite my fingers off.

When the kids were toddlers Zack had been relentless, chasing them from room to room and biting them whenever he caught them. "Get rid of that bird!" John said finally, exasperated.

"No way!" I shouted furiously. "Zack was here first!" I ran interference between the kids and the outraged macaw until Mac turned five, when for no apparent reason Zack's world started to revolve around one of the children he had previously wished only to maim. He treated Mac like a human perch, climbing up his leg, clinging to his belt, and riding on his shoulder; when Mac sat down Zack snuggled under his chin, fluffing out his feathers and grinding his beak in contentment. In the space of a day I went from Zack's favorite person to his mortal enemy, from the loyal owner of a strong-willed but loving bird to the despised jailer of a homicidal mental patient.

"This is why parrots shouldn't be pets," I'd say, removing the hysterically protesting macaw from Mac's shoulder by sandwiching him between two thick oven mitts. "Only one person in a million can put up with them."

"Don't remind me," said John.

Zack had softened over the last few years and would cozy up to me as long as Mac was not around. Taking him away from Mac with bare hands, however, was still out of the question.

Skye appeared, covered with dirt, at the sliding glass door. "I just built another room onto the kelpie house," she announced.

Skye's world was filled with fairies. Her own personal fairy was Marigoldy, who, on certain magical nights, would write tiny notes in tiny handwriting and leave them, colorfully illustrated and accompanied by a real marigold, under Skye's pillow as she slept. Marigoldy's friends were the fairies of the clouds, the sun, the rain, the hemlock trees, and every natural wonder, and each one would eventually leave a tiny letter containing a self-portrait, news from the fairy world, and helpful hints on how to deal with the trials and tribulations of first grade. Sometimes the fairy idea well would run dry and suddenly all the fairies would decamp to the moon or the bottom of the ocean, leaving a farewell note saying "Back in a month" and "Bye! We'll miss you!" Skye would retreat mournfully to the backyard and lavish her energy on the kelpie house, which she was carefully constructing from flat rocks and various pieces of forest flotsam.

Telling the kids about kelpies seemed to me to be a fun way to pass the time and to pay homage to my Scottish ancestors; it was only later, when I watched Skye describe them to a group of friends and their mothers, that I realized why the Brothers Grimm had lost their popularity.

"Kelpies are little men," she said to her wide-eyed audience. "They're thousands of years old. Usually they live in lochs in Scotland, but we have one in our pond and his name is Donal MacLeod. If you make friends with him, he'll tell you all the secrets of the universe. But he keeps his pearls at the bottom of the pond, and if you try to steal them he'll appear as a giant black horse who is so beautiful you have to climb onto his back, and the second you do, WHAM! He'll rear up into the air, screaming his rage, and then he'll drag you down deeper and deeper into the pond until your lungs fill up with water and you drown! And then your hair will turn to seaweed and the crabs will eat out your eyes!"

Skye was delighted, the kids were awestruck, but the mothers' frozen smiles indicated that my Parent-o-Meter had once again fallen to zero.

"Are you ready for the worms?" I said to the kids. "Mac, you get the fish tank and Skye, you get the food."

Mealworms need a healthy diet or they won't do the birds any good, so they are fed a basic mixture of puppy chow, avian vitamins, and several other ingredients that vary according to each rehabilitator. Waxworms are kept refrigerated in the containers in which they arrive; they can go for so long without food that by the time they require it, they have normally already become a food source themselves. We set everything up on the table on the deck, then went to work.

After the fish tank received two inches of mixed worm food, I opened a mealworm bag and slid the crumpled newspapers into the tank. Each time I unfolded a section hundreds of mealworms slid downward, coming to rest in a wiggling, squirming heap.

"Aaaaaggghhhhh!" shrieked the kids, shuddering gleefully and taking turns holding overflowing handfuls of worms.

We opened a container of waxworms so we could scrutinize them, then picked up a few individuals and held them in our hands. Softer and more delicate than mealworms, the waxworms elicited a more restrained response: gentle pokes, followed by long, drawn-out hisses and violently contorted expressions.

"What if Daddy eats them by mistake?" asked Skye, watching as I tucked the four light blue containers into the back of the refrigerator.

"We'll have to warn him when he gets home," I replied, chopping up a carrot and an apple and tossing the pieces into the fish tank. "Meanwhile, where should we put the mealworms?"

"They can stay in my room," offered Mac.

The doorbell rang. "That's a friend of Maggie's named Jen," I said. "She's bringing us two common grackles. Come on, we'll just stick the tank in the dining room for the time being, and deal with it later."

Soon the three of us were walking out to the flight cage with our newest arrivals. I carried two cardboard boxes, while Mac and Skye toted food and water.

"Now remember what we talked about," I said when we were all inside. "Even though these guys are imprinted, they're still wild birds."

"Wild birds think of us as huge predators," said Mac.

"We have to move very slowly and not stare at them," said Skye.

"They'll probably be awfully scared," said Mac.

"Exactly," I said, and opened the boxes. Two dark birds hopped out, surveyed the situation with startling yellow eyes, then one flew onto my arm and the other onto my head. I looked up to see the kids staring at me reproachfully.

"They don't look very scared to me," said Skye.

The grackles were about two weeks apart in age, still dressed in the brownish black plumage of juveniles. They were clearly surprised to see each other,

and equally surprised to find themselves in a 200-square-foot enclosure filled with trees and leafy branches. Within a few moments they were exploring their new surrounding, all the while keeping a close eye on each other. We filled a large shallow dish with water, arranged an appetizing plate of moistened puppy chow, grapes, hardboiled egg yolk, pasta, and live mealworms, then left them alone in the flight cage.

On the way back to the house we stopped and piled onto the hammock, swinging gently back and forth and enjoying what was left of the sunny and peaceful spring day.

"Oh, my God!" came a bellow from the house. "There are maggots in the refrigerator!"

"You're in trouble again," said Skye.

"Wait till he gets to the dining room," said Mac.

Chapter 10

TWEEZERS

I had sworn not to take baby songbirds.

The general public tend to be impressed by those who care for big, aggressive birds: swans, who can break your arm with one wing, or herons, who will occasionally try to stab their beak through your eye, or great horned owls, famous for the fly-by scalping, which is the avian version of the drive-by shooting.

Those birds are a piece of cake compared to baby songbirds.

Tiny, delicate, and insatiably hungry, baby songbirds are food-processing machines. When they're hatchlings (just born) and young nestlings (older but still unfeathered), they need to be fed every fifteen to twenty minutes from sunup to sundown. Then they knock off for the night, giving whatever exhausted creature is caring for them—be it avian or human—a little time to collapse before work resumes at daybreak.

When the babies' pinfeathers start coming in the feedings can be moved up to every half hour, then the time between feedings can be slowly increased in increments of five minutes. When they're around 2 1/2 weeks old, their feathers have opened and they're out of the nest and perching, and you're practically on vacation—feeding them only once an hour.

Since I had two kids and a limited amount of time, raising baby songbirds was simply out of the question. But then the phone rang.

"Suzie," said the woman on the phone, her voice shaking. "This is Liz—do you remember me? Dana's friend? I have a nestful of baby blue jays. I've called everywhere and I can't get anyone to take them and they're hungry and I'm afraid they're all going to die."

"Are you sure the parents aren't around?" I asked. "How long has it been since you've seen them?"

"The mother was hit by a car," she said. "I saw it happen. They've been alone for two hours and I haven't seen any other blue jay go near them."

"I'm not set up for babies," I said. "Let me make some calls."

"Can I bring them to you while you're calling?" she said. "They're all falling over and I don't think they have much time left."

I hung up and immediately dialed Maggie's work number. "Maggie!" I said. "I have a nestful of blue jays coming in. What do I do, besides get this woman to drive them down to you?"

"I can't take them!" whispered Maggie. "We're getting reviewed this week and there are people all over the place. I have nine babies in three nests, and they're all hidden in my desk drawers and if anybody finds them I'm going to get fired!"

"But you have to take them!" I said. "What am I supposed to do with baby blue jays?"

"Call Joanne," whispered Maggie. "Meanwhile, get them hydrated with drops of Pedialyte and feed them mealworms and that dip I gave you. Somebody's coming—I gotta go!"

Cursing under my breath, I called Joanne. No answer. I had the numbers of a few other rehabbers, all a little over an hour away. Nothing. Finally I called Jayne Amico, the Connecticut songbird guru who, at any given spring or summer moment, can have forty to fifty nestling songbirds going at one time.

"Jayne!" I said into her phone machine. "Pick up the phone! You gotta help me!"

Jayne lifted the receiver. "Damn those raptors!" she exclaimed. "I hate those things! How can you rehab them? I've got a Cooper's hawk hanging around my backyard and I know he's going to get one of my little woodpeckers as soon as I let them go."

"Forget the raptors," I said. "Blue jays—I've got baby blue jays and I don't do babies."

"Oh, yes you do!" she chortled. "You do now, honey!"

Technically, blue jays aren't even songbirds—they're Corvids, the group of birds that also includes magpies, crows, and ravens. As Jayne explained as she was giving me a crash course in housing and caring for orphaned songbirds, I was lucky that I was starting out with a nestful of relatively sturdy birds; they could have been impossibly minuscule creatures like wrens or chickadees.

They arrived in their own large and beautifully constructed nest, six awkward, naked hatchlings sprouting tufts of down. They had oversized square heads, bright red mouths, and yellow gape flanges—the outer lining of the mouth, which is one of the markers for identifying nestling birds. Their eyes were just beginning to open, which meant they were about three days old. By the time they arrived they had missed more than ten feedings and were lying limply, like small plants that had been deprived of water.

First they needed to be warmed up, which meant placing them on a heating pad covered by a thick cotton towel. Meanwhile I twisted a small cotton towel into a doughnut, draped another one over the top of it, covered it with a few Kleenexes, and placed the whole thing into a ceramic bowl, creating a clean— and easily cleanable—nest. When they were warm I transferred them to the new nest and rehydrated them by placing tiny drops of electrolyte solution along the sides of their closed beaks until they were alert. Then I began to feed them, something I would do about a gazillion times during the next month.

The best diet to feed orphaned passerines is another contentious issue among rehabbers, inspiring lively bouts of namecalling and slander. A rehabber's goal is to mimic as closely as possible what the parent birds offer their nestlings, and almost all passerines feed their babies bugs. However, the parents supplement with various other wild foods—plus the adults' saliva contains essential nutrients, all of which you must attempt to duplicate if you want the baby to grow up healthy. This means you must put together a complicated and carefully measured vitamin mixture—of which there are many recipes that are constantly being perfected—and puree it into a hummus-like paste, into which you dip each bug and then serve it using tweezers or forceps. At least, that was the idea at that particular point in time. Jayne recently told me that she has jettisoned the dip idea in favor of a more complicated mealworm diet—something that might have saved me hours of work had I known about it back then.

I gratefully defrosted Maggie's container of dip, which she had insisted I take "just in case." I took a portion of it and added a bit of water, cut a group of mealworms in half, and went to work. But as I discovered, nestling birds who have been yanked away from their parents and placed in a bizarre new environment don't automatically open their beaks at the sight of a pair of tweezers.

"Jayne!" I shouted through her phone machine. "What am I supposed to do now?"

The phone clicked on. "Who is this?" she demanded. "Could it be the former raptor rehabber who has finally started to see the light?"

Armed with Jayne's arsenal of tricks I gently tapped the sides of the orphans' beaks, stroked the sides of their faces, lightly jostled their nest, and approached them with tweezers five different ways; soon they were all gaping for food except the smallest one, who steadfastly refused to open his beak. Eventually I pried it open gently, using a tiny tool so as not to inflict any damage, and placed bits of food inside. Realizing that I would be doing this every twenty minutes, rain or shine, I suddenly understood why most rehabbers react badly when yet another mom telephones and announces that she's found a baby bird and wants to know if her five-year-old can raise it.

By the time the kids returned from school things were manageable, if not under control. I had put the nest bowl into a small lidless cardboard box, just to be safe; the kids peered over its edge and gasped.

"Ohhhhh, look at them," cooed Skye, who, from that moment on, would always turn maternal at the sight of a nestling. "Can I help you feed them?"

"Can you find out who hit the mother?" asked Mac, who had recently reacted to a schoolmate's tale of removing an egg-filled nest from a tree by loudly announcing, "You've just violated the Migratory Bird Act—one phone call from me and you're headed for jail!"

"Sorry, Mac," I said. "I'm afraid whoever hit her is long gone."

The kids had just watched their new Harry Potter video several hundred times, so they christened the orphans Harry, Ron, Hagrid, Norbert, Professor McGonagall, and Albus Dumbledore. (Skye had decided, inexplicably, that she was saving the name Hermione for a woodpecker.) After several sessions of watching me feed them, the kids took supervised turns. We gave extra care and extra feedings to Albus Dumbledore, the smallest, while I tried to prepare them for one possible outcome.

"He's so little," I said. "It's like being the runt of the litter. He's just not as healthy as the others. Sometimes not all of them make it."

"He'll make it," said Skye.

The odds are that not all of them would have survived in the wild. The largest and most aggressive siblings usually get most of the food, while the smallest become progressively weaker and sometimes die. Occasionally the parents will push a sickly baby from the nest, an act that may seem brutal to those who don't understand the Herculean task the parent birds face. Once you stop to consider the dawn-to-dusk feeding schedule and combine it with the dangers facing most songbirds—both natural (natural predators, bad weather) and man-made (suburban development, outdoor cats, windows, cars, pesticides, etc., etc., etc.)—it's easier to comprehend a parent bird's cutting its losses early and devoting its limited resources to the nestlings more likely to grow to adulthood.

Our extra labor didn't do any good. Despite our best efforts, Albus remained small and sickly and died two days later. I carried his tiny body into the woods, once again trying to figure out what I would tell the kids when they returned from school. Unlike with the house sparrow, they had invested time, effort, and emotion in the nestling jay, even if it was only two days' worth.

Wildlife rehabilitators see more death in a busy month than most people do in a lifetime, and must come up with their own coping mechanisms. In my previous eleven years I had handled the deaths of many wild creatures with the emotions rehabbers strive for: a mixture of regret and resignation and a resolve to use any knowledge gained for the next one. But I had one spectacular crash. She was one of a pair of orphaned crows I had raised during my years at the raptor center. I released them both and she stayed around the house, only to die in a freak accident one late summer morning. She was half wild, still friendly to our family, but along with her shyer nestmate, she was in the process of forging a bond with the local crow flock. She soared between our world and theirs, bursting with life and joy, and when she flew beside me as I ran through the woods I felt as if I, too, were flying.

I knew I could lose her at any moment. Like the chickadees of my childhood she was free to leave, free to cast off the chains of my increasingly desperate love for her. Though captive-raised, she was my tangible link with the wild world, the feathered embodiment of everything I had always found wondrous but unattainable. I rejoiced when she appeared and feared for her safety when she left. I worried about a hawk attack, however, not some random, unpreventable accident: a collision with a swing that broke her neck.

After she died I swore to myself that I could still see her flying beside me as I ran through the woods, and grieved for her for months with an intensity that frightened everyone but my children. Wearing an unconvincing smile, my eyes bruised and swollen, I would start them on a project; as soon as they were engrossed I would slip out the door, hurry down the hill, and sit beside the stone-circled grave blanketed with flowers, my face buried in my arms. Soon they would both appear behind me, at ages four and five the small guardians of their devastated mother. "Time to come home now," they would say, and carefully lead me back up the hill to the house.

Remembering this I realized that they were more resilient than I gave them credit for, and probably far better equipped than I was to handle the highs and lows of bird rehabilitation. The kids came home from school, peered into the box, and looked up in dismay.

"Where . . ." Skye began.

"I'm so sorry," I said. "He didn't make it."

"Oh, no," sighed Mac.

For a long moment Skye stared into my eyes, precariously balanced between grief and resignation.

"Could you feed the others?" I asked her.

Her gaze dropped to the remaining five nestlings. Roused by our voices, they had lifted their heads and were opening their beaks.

"Okay," she said finally, and picked up the tweezers.

Chapter 11
BACKING TOWARD THE CLIFF

My favorite description of the grackle comes from Ogden Nash:

The grackle's voice is less than mellow,
His heart is black, his eye is yellow,
He bullies more attractive birds
With hoodlum deeds and vulgar words,
And should a human interfere,
Attacks that human in the rear.
I cannot help but deem the grackle
An ornithological debacle.

From what I could see, our grackles were having a fine time in their flight cage. The younger one was somewhat tentative, the older already swaggering. Gender was anyone's guess, so I declared the younger one female and the older, male. Both had a repertoire of chirps, clicks, and that endearing grackle sound of heavy metal being dragged across concrete. Although they were still young, you could already catch glimpses of the fearless, aggressive personality of the adult grackle. We christened them—in order of ascending age—Null and Void.

During their first two weeks I visited them twice a day, keeping our interactions to a minimum, hoping they would come to view me as a relatively boring food source rather than a parental figure or a fun pal. I combed through my ever-expanding bookshelves for specific information on readying captive-raised grackles for release, surfed the Net, and peppered my wildlife rehab electronic mailing list with questions.

According to the veterans who had released captive-raised songbirds, the two most important release criteria were the bird's ability to relate to its own species and its ability to forage for its natural food. Slowly but surely the grackles were beginning to interact more with each other, although Void occasionally acted irritated with Null, like a cool kid with a bothersome younger sister. The "natural food" part took a bit more doing. I looked up "Grackle, Common" in Marcy Rule's *Songbird Diet Index*, one of the songbird rehabilitator's bibles, which contains both the natural and captive diets of well over 100 species of songbirds. Armed with clippers and gloves I scoured the countryside for various berries, weeds, and grasses gone to seed; brought them home; and arranged them artfully around the flight to encourage the grackles to forage. Meanwhile, the kids and their friends took to the fields with buckets and bug boxes and returned with beetles, grubs, caterpillars, crickets, and centipedes. The first time they dumped a bucket of bugs onto the flight floor the grackles danced backward, their yellow eyes bright with alarm, as they were used to their relatively slow-moving mealworms and not things that hopped and raced across the ground. Soon, however, they were outmaneuvering the crickets and expertly turning over leaves and wood chips to find the centipedes hidden underneath.

The wound under the house finch's eye had healed and he was back in the second flight cage, keeping company with a song sparrow recovering from a broken wing. The robin who had lost the bird fight was still healing, his prodigious appetite no doubt fueled by dreams of revenge. My plan to give the blue jays to Joanne was loudly vetoed by the kids, who had decided that raising baby birds was a grand and worthwhile project, and seconded by Joanne, who was swamped with a dozen or so nestlings of her own.

The five remaining nestlings had rallied and were eating like champs, although Norbert, the second smallest, developed foot problems and needed snowshoes. A "snowshoe"—used to straighten out a bird's foot when it is curled and clenched—is made by affixing a piece of lightweight padded plastic, made to order for each bird, onto the bird's open foot using special tape. Three days was sufficient to straighten out Norbert's toes, but meanwhile one of the others had developed bloody diarrhea. A fecal test revealed protozoa, so the whole crew had to be wormed.

And then Jill called.

Jill Doornick is the founder of both Animal Nation, an animal rights and rescue operation, and Westchester's Wildlife Line (WWL), a group of rehabbers based in the county south of me. When people call the WWL, they choose from a menu of wildlife forms ("For small mammals, press 1. For reptiles and amphibians, press 2 . . . "). They then hear a recorded message telling them what to do and whom to call.

"There are no secrets around here," said Jill. "I've heard all about you and I know you have two beautiful flight cages. What I'm hoping is that you'll let me list you on the hotline. We don't have enough bird rehabbers and everyone is always swamped, and it would be such an incredible help to all of us."

"I'm sorry," I said. "I have the two flight cages, and I'm happy to take any birds who need them. But I can't take injured birds. I don't have the space and I have two kids in elementary school."

"That's okay!" said Jill enthusiastically. "My kids used to help me rehab! Can you do babies? They don't take up much space. A lot of rehabbers work during the day and can't take nestlings to work. Are you home during the day?"

"Yes," I said, "but the kids get home at three."

"Do they like birds? Can they help you feed babies?"

"Well, as a matter of fact we have five blue jays, but that was just an accident and there's no way . . ."

"There! See? You can't imagine what having you on board would do. We get so many calls for these poor birds, and we just don't have enough people to take them, especially in your area. It's awful."

"I understand that," I said. "But . . ."

"I'll tell you what," she said. "Could you just provide information? Tell people if they've found a baby bird to put it back in the nest, tell them to leave the fledglings alone—you know, that kind of thing. That way the other rehabbers won't have so many messages backed up. And if it's really a rescue situation and you can't do it, you just tell them to call someone else."

I considered it. How could I say no to just giving out information?

"Please help us," she said. "We really, really need help."

"All right," I said.

I hung up the phone, wondering what sort of pact with the devil I had just made. I needed to take a run and think things through. I glanced down at my weekly calendar and saw, to my delight, the word *goshawks* scribbled in purple magic marker.

By now the goshawks' eggs would have hatched and the nest would be filled with nestlings. Since our audio encounter her voice had haunted me, echoing through my dreams, winding through my head like an old song I couldn't dislodge. I fed the blue jays, then donned my running clothes—shorts and an old tee-shirt riddled with parrot beak holes—and added a pair of ski goggles, just in case I met up with one of the parents and they were not happy to see me. I pulled the goggles over my head and around my neck, grabbed a small pair of binoculars, and took off into the woods.

The June woods were welcoming, filled with dappled sunlight and the sounds of summer, and the farther I ran the better I felt about agreeing to list my number on Jill's phone bank. I still had control over the situation; I could provide information to the public, which I was more than willing to do, and say no to anything that would upset the household equilibrium. Feeling confident and decisive, I slowed down as I neared the nest area.

Slightly smaller than the more familiar red-tailed hawk, the adult northern goshawk has a slate gray back and wings, a pale gray chest, and a dark band of war paint across its eyes. Like many other raptors, the female is noticeably larger and heavier than her more agile mate, which allows them to hunt prey of different sizes—an advantage when there are hungry nestlings to feed. The female is also louder and, around her nest, more aggressive.

I dropped down to a walk, squinting through the slight camouflage provided by hemlocks almost stripped of their greenery, and found the nest filled with whipped cream. At least that's what it looked like at first: a soft white froth atop a shadowy structure, lightly trembling and shifting until it suddenly revealed the dark eyes of a nestling goshawk. There was another beside the first, maybe even two, but I was too far away to see them clearly. I was so transfixed that several seconds must have passed before I realized that the male goshawk was standing on the edge of the nest, staring at me intently. I backed away slowly, hoping if I put some distance between us I could watch them for a minute or two, then leave them all in peace. I continued until I came to a fallen tree, then quietly sat down and raised my binoculars.

As soon as I started to focus on the nest I felt a strange sensation. It was a slight discomfort, a shiver, a feeling that made me think, "Uh-oh," although at that moment I was too slow-witted to figure out why. I lowered my binoculars and there, at eye-level and not ten feet away, was the female goshawk.

When predators come for you, you know it. And they know you know it.

Like a snake before a snake charmer I sat dazzled, held captive by her un-wavering red eyes. I have no idea how long we stared at each other. Finally I did something that was logical for a birder, but wildly illogical for a birder sitting a few feet away from a hormonal raptor: I raised my binoculars.

I suspect that from the goshawk's point of view, raising the glasses to my eyes not only made me even uglier and more misshapen than I was before, but raised my threat status from code orange to code red. In any case, she launched herself forward off her branch just as I launched myself backward off my log.

She flew over my head and landed screaming on a high branch, her voice ringing through the woods: *kek-kek-kek-kek-kek-kek-kek!* I backed away, keeping in mind that while a great horned owl might be *capable* of doing more damage to its chosen target, a northern goshawk is *willing* to do more. Once again, she hurled herself straight for my head. When she was several feet away I dropped to the ground, and she banked, headed up, and landed on a tree limb. I had dodged her first bullets not because of any vast raptor experience but because of my hardwired instinct for self-preservation; had I any control over my actions I would probably have simply stood there gaping, then shouted, "Oh, wow! Do you have any idea how cool you are?"

I continued to back away, hoping against hope that her mate wouldn't decide to join the assault, and I put a large, leafy maple between us. In response she leaped into the air, pumped her wings once, turned sideways, flattened her body, snaked around the tree trunk, and came sailing through the leaves at my face, knifing through the air like a shark through water, a sleek and vengeful Fury with a single goal. I ducked, she banked. We performed our duet twice more as I continued to retreat and she to chant her outrage, until she had driven me from her territory and banished me from her sight.

Safely out of range I did a victory dance of sheer exhilaration, suddenly understanding the psyche of the disciple who waits for hours in the rain for a quick glimpse of the spiritual master. Far from the world that humans have reduced to tree stumps and pavement, the goshawk showed me a kingdom where my species gave me no unfair advantage, where I could see a fearless wild creature in all its glory, where I could participate in an ancient ritual that ended in a fair and just way. I turned in a slow circle and saw a nuthatch work its way up an old oak tree, two titmice peer at me from a striped maple, and a thrush sail by and disappear into the deeper woods. From the distance came the ascending echo of a pileated woodpecker. Theirs was the world in which I longed to play a part, in which I could witness the transcendent and perhaps even begin to atone for the sins of my own kind.

I raced back home, the forgotten goggles clutched in my hand.

Chapter 12

DAYCARE

"Rrrrrrrrrrrrr," growled Skye softly.

Holding the bug-filled tweezers above her head, she slowly serpentined them down to a tiny eastern towhee, a new arrival who had steadfastly refused to eat. "Rrrrrrrrrrrrrr," she repeated, and to my amazement, the nestling promptly opened its beak.

"Airplane," she explained. "Remember, you used to get me to eat by pretending the food was an airplane."

The kids were out of school and day camp had yet to start, but those carefree summer days were now chopped into twenty- to thirty-minute segments. People would call to say they had found a nestling bird on the ground, and every once in a while I would succeed in getting them to climb into a tree or up the side of their house and put it back in its nest. More often than not there were complications: the nestling had an obvious fracture, the nest was forty feet up, or the bird had appeared out of thin air and there was no nest in the vicinity. If it was a weekend, I could occasionally get the finder to call Maggie or Joanne; more often than not, I would simply take it myself. The nestling would arrive, receive the standard-issue box and towel nest, and somehow join the schedule. I pressed the kids into service, and Skye became a master at getting reluctant babies to eat.

Nagged by guilt that I was neglecting my own children, I'd pack the nestlings into a big wicker picnic basket with a hinged (thus closeable) top, place their food and supplies into a carryall, put the kids' stuff into a giant canvas bag, then drive to the pool or the lacrosse game. Mac and Skye would race ahead and I'd stagger after them, laden like a pack mule and muttering under my breath, then I'd set up shop under a tree while they expertly fielded questions from the circle of kids who would inevitably gather around us.

"They're blue jays. Those are finches, and that one's a cardinal. They eat bugs and dip. They eat all the time. No, you can't raise one by yourself; you have to have a special license to raise a wild bird. You once gave a baby bird bread and milk? Birds aren't mammals; they don't drink milk! No, don't *ever* give them water; if a baby bird opens his mouth and you squirt water in, you'll drown him. If you find a baby bird who fell out of his nest, just put him back; the mother bird won't care if you've touched him, she just wants her baby back."

Fledglings were a different story. Fledglings are fully feathered and ready to leave the nest, but they haven't gotten the hang of flying and usually are still being fed by their parents. Often called "branchers," they hop from branch to branch, occasionally missing their target and falling to the ground. In a perfect world this would not be a problem; they would simply hop around until they found a bush or limb to scramble up, and life would continue. However, thanks to the world of humans, when the fledgling falls to the ground it often encounters a cat, a dog, or a child.

Some of the people who found me through the wildlife hotline couldn't have been more helpful and concerned. "I can see the parents flying around," they would say. "I've put my dog inside, and I've told the kids not to go near that area. What else can I do?"

"You did everything right," I'd say. "Wait for an hour and check on him. If he isn't gone, then pick him up and put him on the highest branch you can reach. Then just leave him alone and let the parents take care of him."

Others made me want to reach through the phone and grab them by the throat. "Oh, I couldn't possibly bring Mr. Whiskers inside," they'd say. "He'd be mad at me. Besides, everyone in the neighborhood has cats so if Mr. Whiskers doesn't get him, somebody else will."

I tried mightily to find other rehabbers to pawn the fledglings off on, but there simply weren't enough of us. That first summer I ended up with a collection of fledgling robins, which meant that each morning I could be found bent double, bucket in hand, combing through dead leaves for earthworms.

"I'll pay you," I finally said to the kids. "Fill up these containers and I'll give you each a whole dollar."

"A dollar!" they said scornfully. "Is that all?" Disappearing into the woods, they quickly reappeared and handed me back the containers. I removed one of the lids, revealing a half a dozen of the biggest earthworms I'd ever seen.

"Jeez Louise!" I exclaimed. "Those are pythons! I want the robins to eat the worms, not vice versa."

"Up to you," said Mac, shrugging. "We can get smaller worms, but it's gonna cost you."

Wild birds can't be kept in regular birdcages, as they will damage their feathers by brushing them against the metal bars. To supplement my collection of heavy plastic pet carriers I bought two reptariums—roomy reptile enclosures made of light plastic frames surrounded by soft mesh. Filled with leaves and branches, they provided a safe and recognizable habitat for fledglings who had suddenly been taken from their own environment.

Young birds grow with astonishing speed, and there were mornings when I would peer into a nestful, marvel at their change from the previous evening, and feel lucky and privileged. The kids took their assistant duties seriously and were thrilled with their newly earned right to feed the birds without supervision. Occasionally, however, our inability to locate our dog-eared copy of the nestling identification book caused our conversations to spiral into absurd proportions.

ME: Yup, it's cute all right, but I have no idea what it is.
MAC: Maybe it's some kind of warbler.

SKYE: Maybe it's a woodpecker.

ME: Maybe it's an elegant trogon.

MAC: I think it's a blue-footed booby.

SKYE: Nah . . . it's a dodo.

ME: It must be some kind of strange dwarf condor!

MAC: It's a miniature featherless moa!

SKYE: It's half stork, half beaver!

ME: Let's try feeding it an enchilada suiza!

MAC: And some Cheez Doodles!

SKYE: I think it wants a chocolate sundae with lots of sprinkles!

How many kids even know what a moa is? I thought triumphantly, somehow convincing myself that my children's ability to identify an extinct New Zealand ratite was far more important to them than having a mother with a decent amount of free time. The days were hectic but sometimes went fairly smoothly, leaving me with outsized feelings of capability.

Other days did not go as smoothly, leaving me with outsized feelings of inadequacy. As any rehabber will attest, problems tend to come in clusters. One bird will suddenly stop eating, another will develop a mysterious limp, a third will look a little "funny," and then the telephone will start ringing.

"Can you take a single baby mallard?" asked the voice on the phone.

This was one of the days that were not going smoothly, and I was feeling rather peevish. "A duck!" I burst out. "You must be kidding me! I'm up to my eyeballs in blue jays and robins, and God knows what else, and I'm not even supposed to be doing babies! Out of all the birds I don't do, the ones I don't do the most are ducks!"

"Listen," said the voice. "I must have called every rehabilawhateveritis in the state, and either they tell me they're full or they don't answer their phone. We're leaving for the airport in two hours and if I don't find somebody to take this duck we're going to have to just dump him back off where we found him."

"Take him back to his parents!" I said.

"We couldn't find them. He was running around by himself. We looked all over for the parents—believe me, we looked *everywhere*."

"I don't have any duck food!"

"There is a feed store two miles from here. I will go in and buy the biggest bag of baby duck food they have. I can deliver the duck and the food right to your door; all you have to do is tell me where you live."

"No way!"

"Please!"

"No!"

"Please! If you don't take him he's going to die! *Please! You can't let him die!"*

Eventually I hung up the phone. " 'I don't do ducks! I don't have any duck food!' " I chanted nasally, mimicking my pathetic self as I trudged off to my purple three-ring binder to look up "Orphaned Mallards."

The beauty of baby ducks is that they feed themselves. Unlike altricial songbirds and raptors, whose babies are born blind and nestbound, ducks are precocial; that is, 24 to 48 hours after their birth they're wide-eyed and racing around after their mothers. This made me feel better, as did the fact that the incoming duckling wasn't a wood duck. I had learned through the grapevine that most rehabbers live in fear of getting wood ducks, who are so shy and reclusive that they drop dead from sheer terror if you so much as glance in their general direction.

I went off to prepare a duck box, suddenly filled with gratitude toward the rehabber friend who had insisted on giving me a feather duster "just in case," even though at the time I had insisted that there was no way I would ever need one. A duck box is a large shoebox with the lid attached on one side, allowing it to be easily opened and closed. An entrance hole is cut into the front. Another hole is cut in the top (the ceiling), into which is stuck an old-fashioned feather duster, so the feathers are in the box and the handle sticks out of the ceiling. A heating pad covered with a terry cloth towel is placed on the bottom, then the whole contraption is placed into a large topless cardboard box. The end result is a nice roomy enclosure, complete with a little house the duckling can run into if it's cold or scared. The heating pad provides warmth, the feathers are an approximation of a mother duck. I put the whole thing on the washing machine, closing the folding doors just as the front doorbell rang.

A young couple stood outside the door, both wearing apologetic but determined expressions. The woman held a small canvas carryall, the man a very large bag of Unmedicated Game Bird Starter. Wordlessly, the woman opened the carryall.

I looked in. Staring back at me with an apprehensive expression was the tiniest duck I'd ever seen.

"Oh God," I croaked. "Are you sure you looked everywhere for the parents? *Everywhere?*"

"I swear to you," said the man, solemnly raising his right hand. "We looked *everywhere.*"

I carried the duckling into the kitchen, where John and the kids had materialized. The tiny duckling's appearance caused a waterfall of elongated vowel sounds, with only one family member withholding approval.

"Aaaaaaaaaah!" breathed Skye.

"Oooooooohh!" sighed Mac.

"Awwww!" crooned John.

"War!" shouted Mario.

We all looked at each other.

"I've never heard him say that before," said Mac.

"Maybe African greys don't like ducks," said Skye.

I put the duckling in the box and gave it a small tray of soaked food, which, after some encouragement, it ate eagerly. Afterward I opened the lid of the house, placed the duckling on the nice warm towel, and fluffed the feather duster around it, then closed the lid. There was a moment of silence, then a soft peeping.

"I'm going to name her Daisy," said Skye.

"Daisy!" snorted Mac. "No way."

"What would you call her?" Skye demanded. "Brutus," she intoned deeply. "Lothar."

"You know something?" I asked. "I'm not sure that's a mallard. I think it looks kinda funny."

"What do you mean, 'kinda funny'?" asked John.

"It's a scientific term," said Mac. "I'll explain it to you when you're older."

The peeping grew in volume and intensity. Soon the peeps were accompanied by soft thuds as the newly christened Daisy began attempting to hurl herself out of the box. From our various positions we could see the top of her head appear, then disappear, then appear again. Finally she levitated straight up and somehow landed on the rim of the box, where she teetered wildly before pitching off toward the bare floor five feet below. Before I knew what I was doing I'd thrown myself across the room like a star football receiver, landing flat on the floor and catching the duckling just before she hit the ground. I lay there silently, wondering if I'd broken any bones.

"Nice one, Mom!" came the appreciative chorus. "Do it again!"

Soon the duckling was back in the box. The box now sported a top made of metal hardware cloth, the same material that encased the flight cage. As I listened from the living room I could hear Daisy's levitation efforts being thwarted by the wire top. Bonk! Peep, peep, peep. Bonk! Bonk!

"Awwwwww, she's lonely!" called Skye, picking up the ringing telephone and disappearing into her room.

She was lonely, all right, but it was seven at night and I didn't have any other ducks to keep her company. I covered the box with a towel, thinking the darkness might calm her down and make her sleepy. No dice. I continued to listen to the steady drumbeat of Daisy's head against the top of the box, desperately hoping she'd settle down before she gave herself a concussion or died of stress. Bonk!

I returned to my bookshelf but found no information on what to do if your single duckling adamantly rejects the duck box you have so painstakingly put together. My instinct was to comfort her and deal with imprinting later, but then I envisioned Daisy as a mature duck irreparably bonded to humans, living in a miserable gray half-world, a pet but not a pet, a wild duck but not a wild duck, her dire fate inarguably my fault. I'll just leave her in the box, I thought, then envisioned the tiny, terrified duckling bereft and abandoned, finally dying a lonely death from the large dent she'd put in her own head. Either way I was headed for Rehabber Hell. Rehabber Hell materializes around your bed at two in the morning, when you lie awake and catalogue—bird by bird—all the rehabbing mistakes you've ever made, both real and imagined, and inevitably conclude that you are nothing but a drunk driver careening wildly through the helpless bird community. This is one reason rehabbers always look so tired, even during the off season.

Finally I couldn't take it anymore. Knowing I was destined for Rehabber Hell no matter what I did, I strode into the kitchen, scooped up the frantic duckling, and marched back into the living room, where Mac sat engrossed in *The Seven Songs of Merlin*. I deposited Daisy under his chin. Without taking his eyes off his book he cradled one hand around her, and Daisy nestled in with a blissful sigh and instantly fell asleep.

A half hour later Mac deposited the groggy duckling in her feather-duster bedroom and silence ensued. Perhaps, I thought, this was all she needed to get to sleep. I'd find another duckling for her tomorrow, she wouldn't imprint, and I'd be saved from eternal damnation. Feeling very pleased with myself I readied the kids for bed, then sat down with John to watch *The Sopranos*.

During its heyday we considered *The Sopranos* sacrosanct. Most television is so awful that the Mafia drama was one of our only forms of serial entertainment. John and I would go to ridiculous lengths to avoid being out of the house on Sunday nights, and we spent inordinate amounts of time deconstructing the characters' motivations and speculating on the details of their eventual demise. At the stroke of 9:00 on this particular Sunday night we were in our usual positions: draped on the couch, wineglasses in hand, grinning with anticipation. Just as the theme song started to play, however, there was an unexpected accompaniment.

Peep. Peep. Peeppeep. Peeppeeppeep. Peeppeeppeeppeeppeeppeeppeep Bonk! Bonk! BONK!

In that split second a decision had to be made. But factoring in *The Sopranos* made it surprisingly easy. I raced into the kitchen, and before the theme song ended I was back on the couch, wineglass in hand, duckling carefully ensconced under my chin.

With my luck it happened to be the episode in which Christopher (pronounced Chris-tu-fuh), the psychotic young lieutenant, tells Tony, the mob boss, that Christopher's slinky girlfriend Adriana has just admitted to spilling her guts to the Feds. Tony gets his soldier Silvio to drive the unsuspecting Adriana to an upstate hospital, where Christopher is supposedly clinging to life after a baffling suicide attempt. The whole thing is a setup, of course, and before you know it Silvio has pulled off the highway into a deserted wooded area, where Adriana instantly realizes the jig is up.

Now I always had a soft spot for Adriana, who was rather dim but essentially had a good heart. She'd sashay across the screen wearing a catsuit, four-inch heels, and a black eye from her latest encounter with her loutish boyfriend, and I'd wish I could rescue her, put her in my flight cage for a month, and set her free. But it was not to be. As I lay mesmerized on the couch Silvio jumped out of the car, opened the passenger door, threw Adriana to the ground, and pulled out a giant silver gun. Forgetting all about the duck under my chin I sat bolt upright and, embarrassingly enough, actually shouted "Look out!" to the doomed Adriana. This propelled Daisy, suddenly wide awake and peeping hysterically, through the air and into a cushion at the other end of the couch. John leaned forward and picked her up, his eyes still glued to the screen.

"Here," he whispered. "You dropped your duck."

Daisy's sudden launching must have tired her out, for the rest of the night passed uneventfully. First thing in the morning, filled with foreboding, I tiptoed to the box. I opened the lid of the duck house and there she was, definitely alive, snuggled under the wing of a soft toy duck that Skye had donated to the cause. All I had to do now was scare up some more ducklings, and I'd be home free.

Chapter 13

MOMENTUM

"But you're a duck rehabber," I said into the phone. "How can you have no ducks?"

As it turned out, when you don't want a duck they're everywhere, and vice versa.

I spent the morning on the telephone, trying to find a rehabber with potential sibling ducklings. Nothing. I put a message on my wildlife electronic mailing list. "Subject: Any Duck'll Do." No replies. My standards kept dropping. How about a gosling? I'd say. Anybody have any little wild turkeys? Daisy might end up fanning her tail and gobbling, but at least she'd know she was some kind of bird. What about a small chicken?

"I could put her in with my redtail," said one rehabber friend helpfully. "The lion and the lamb, the hawk and the duck. Can't we all just get along?"

In the meantime, I didn't have the heart to leave the fragile, desperate little creature alone. I left her in her box for short periods so she'd get used to it, but otherwise I carted her around in the crook of my elbow. She watched TV with the kids. I let her follow me down the uncarpeted hallway, her frantic footsteps sounding like a stick being dragged across a picket fence at high speed. Once John found me at the kitchen table, lost in thought, Daisy sitting Buddha-like on my lap.

"Don't worry," said John. "You'll find her a buddy."

"I know," I sighed. "It's just that . . . I can't believe they whacked Adriana."

On the third morning we filled the bathtub with warm water, slid Daisy in, and witnessed firsthand the phenomenon we came to call duck joy. ("Duck joy!" said John. "It sounds like something you'd order from a Chinese restaurant.") Daisy gamboled, she shimmied, she dunked her head in and out of the water half a dozen times in a row; she'd rear back, pull her body up, and flailing her legs and wings, high-step across the top of the water in a frenzied version of the Maori haka. Mindful of the dark tales I'd heard of rehabbers who'd left ducklings alone in water for only a few minutes and returned to find them all drowned, I never allowed Daisy to bathe without a spotter; whenever she grew tired or chilled there was always a helping hand to lift her out of the tub and onto a heating pad by a sunny window.

That morning I received a call from Maggie: she had a mallard duckling, maybe a week old, a possible companion for Daisy. "There's only one problem," she said. "He's neurologic."

There were a daunting number of explanations as to why a duckling would be unable to control its movements. Bacterial and viral diseases, environmental poisons, genetic anomalies, a blow to the head. "The woman who found him said she saw the family swimming around on the other side of the pond and they were all fine, so I doubt it's anything contagious," said Maggie. "I took him to the vet and they couldn't find anything wrong. He probably got stepped on. You can try it, and if it doesn't work out I'll take him back."

Maggie's duckling was a half-size bigger than Daisy—an impossibly cute, fuzzy little creature, but one to whom the fates had been especially unkind. He'd lurch from one side of the box to the other and then flip onto his back, where he was unable to right himself. When he was at rest, though, he was fine. He'd look around the box and up at me with an interested expression, and when I held him steady by the food dish he'd eat heartily. I moved Daisy's heated duck house into a cardboard box, rolled up several hand towels, and positioned them at odd angles, like the walls of a maze: when the little mallard flipped onto his back he could push himself up against one of the towels and so get enough leverage to right himself.

He and Daisy were clearly happy to see each other, often ending up in a sleepy little duck pile wedged next to one of the hand towels. I thought the new duck might benefit from water therapy, so twice a day I'd draw them a bath. The first swim was alarming: the new duck spun in a circle, rolled upside down, and needed a quick rescue. But with each subsequent swim he became stronger and more adept, until I started to feel "cautiously optimistic" about his chances of recovery.

When both ducks finally settled into a routine I was ready to tackle the mystery of Daisy's identity, as when they were side by side it was fairly obvious that they were not the same species. I sat down at the computer, typed two words into Google, and closed my eyes. When I opened them there was a perfect picture of Daisy.

She was *Aix sponsa*: a wood duck.

"It's a good thing I didn't know you were a wood duck," I told her later, "or you'd probably be a dead duck."

There are a number of knowledgeable, intrepid souls out there who specialize in wood ducks and have their care down to a science; my friend Wendi Schendel used to raise them for a university-sponsored project in Montana, and would release up to a couple of hundred a year. But for most rehabbers, the orphaned wood duck mortality rate is above 90 percent. On one hand, I felt cool and powerful: Yes! I am Super-Rehabber! On the other, I felt a little like Rosie Ruiz, who had won the 1980 Boston Marathon by hopping onto the subway for the hard part. Sure, I'd won the Wood Duck Marathon, but I'd done it by cheating the system—something I thought about every time I heard Daisy's furious tap dance following me down the hallway.

Then there were Null and Void, who had made great progress in bonding with each other but still hadn't quite kicked the habit of landing on my head. I tried to enter their flight cage and feed them in a brisk and businesslike way, but it was difficult. Filled with standard young-bird joie de vivre, they carried crumpled leaves around in their beaks, stole pebbles from each other, and attacked the hanging ropes; they'd jump up and down in place, energetically flapping their wings until sheer momentum spun them around in a circle, like old-fashioned prop planes. Occasionally I'd relent, reasoning that I didn't want them to *dislike* me; I'd bring them toys, like pinecones and seed pods, or toss a pebble into their water dish, inspiring furiously flaphappy bouts of bathing.

The wing of the adult robin who had lost the bird fight had healed nicely, and he was in the second flight cage with the house finch, the song sparrow, and a small group of fledgling robins who were eating on their own. Although the adult robin did not appear particularly paternal, he did provide a good example for the fledglings by eating, perching, and flying around like a robin. He also taught them how to comb through the bug pit, a square area of deep organic soil held in place by split logs. The kids would dump their containers of small earthworms and miscellaneous bugs into the bug pit, then cover them all

with leaves and grass. As we watched from the outside, the adult robin would stride over and expertly turn over the leaves to find the bugs underneath while his avian students watched in fascination.

There is nothing sweeter and more appealing than a fledgling songbird. They have an air of bright-eyed bewilderment, as if they find the strange new world around them entrancing but slightly confusing. When I first released the young robins into the flight cage, one hopped up on a perch and stood still, regarding her new surroundings with surprise. The house finch quickly joined her, but evidently didn't receive the response he was looking for. The finch hopped to the robin's other side, then behind her, in front of her, and finally onto her head, where he stood briefly before jumping back down to the original perch. The finch did this twice while the robin remained immobile, looking more and more perplexed.

That afternoon Jen Bowman, who had given me one of the grackles, called. "I just wanted to see how the grackle was doing," she said. "And ask if you might have room for a fledgling cedar waxwing."

"A cedar waxwing!" I exclaimed. "How soon can you get him here?"

Cedar waxwings must be among the most beautiful birds in the world. Their buff-colored bodies, regal crests, striking black masks, and brilliantly red and yellow scalloped tail-feather tips all give them a slightly Asian, otherworldly air. Their trilling voices are like tiny bells. They love blueberries. They travel in flocks and land together in trees, the sight of which few bird lovers can ever forget.

Jen's waxwing had been delivered to her as a nestling with a broken leg. Now healed and feathered and eating on his own, he just needed some flight-cage time and, in a perfect world, another waxwing for company. I put him in with the finch, the sparrow, and the robins, entranced by his delicate beauty and praying to the nature gods for another waxwing. Though shy and unsteady at first, within days he was swooping, banking, and turning in midair, learning to forage by combing the flight for the berries I'd painstakingly impaled on dozens of branches each morning. When I entered the flight cage he'd fly to a high perch and look down at me with calm self-assurance, like a small, impeccably gowned emperor from the T'ang Dynasty, while I did my chores.

I tried not to stare at all of them. Prey species (such as songbirds) have eyes on the sides of their heads, giving them a greater range of vision; this way they have a better chance of seeing a predator species, whose eyes are on the front of their heads, coming toward them. In the wild, staring at a bird is a clear signal that you intend to do him serious bodily harm, and I didn't want to cause alarm by steadily watching them all with my predatory human eyes. But I couldn't *not* watch them; each one was so breathtakingly beautiful, every one of their movements so remarkable, that sometimes I ended up facing away and watching them out of the corner of my eye until the resulting headache forced me to stop. Occasionally I would pause, close my eyes, and listen to the sound of their wings. When the sparrow flew by it was like the whispered roll of an Italian *r*, the mysterious half sentence of an overheard conversation. I stored the wing beats of each bird in my memory, each one a quick riff from a different song, so I could replay them all at the end of the day.

I e-mailed my friend Ed and told him about the burgeoning bird population. "Hamlet said, 'there is special providence in the fall of a sparrow,'" he wrote back. "I might add, as well as the fall of a wood duck. And a finch. And a group of jays. And many robins. And now a cedar waxwing. Is there anyone I've missed?"

"Who the heck knows?" I wrote back. "Sometimes I lose track."

It was over two weeks since the blue jays had arrived. They were fuzzily feathered and bouncing out of their nest, so I moved them into a roomy reptarium. The reptarium was filled with tree branches for perching and leaves, pinecones, and acorns for play. One day another rehabber delivered a single jay, only slightly older than ours; he quickly fit into the Harry Potter clan and was christened Albus Dumbledore II, in honor of the smallest nestling who had survived only a few days.

By the third week feeding them was an adventure. They were filled with energy and dying to fly but weren't yet eating on their own, so I couldn't put them in the flight cage. Whenever I'd unzip the top of the reptarium in order to feed them, they'd burst upward like large kernels of blue popcorn and bolt off through the house, with the kids and me galloping behind in hopeless pursuit. This led to breathless exchanges:

JOHN (ENTERING THE HOUSE): Hello everybod—

ME: Close the door! There's been another jailbreak!

JOHN: Not really! How unusual!

ME: Quick—grab Hagrid!

BOTH KIDS (STOPPING DEAD): But that's not Hagrid, it's the Professor.

JOHN: Here, I'll get Harry—

SKYE: That's not Harry, it's Ron.

ME: I've got Norbert!

MAC: That's not Norbert, it's Albus Dumbledore the Second.

Sometimes, after we herded them back into the reptarium, we carried it into my bedroom and placed it on the floor next to the screened door. I put a shallow dish of water on the cage floor and the kids took turns spraying the jays through the mesh with a water-only plant mister, encouraging them to preen and readying them for their eventual encounters with rain. At first they cowered together, frightened by the unfamiliar sensations, but soon they were spreading their wings to catch the drops, fanning their tail feathers, and crowding into the water dish. Later they took up positions on various perches, drowsing in the sunshine that poured through the screen door and listening to the sounds of the outside world.

"You'll be there soon," promised Mac.

Chapter 14

INCLUSION

Breathe in. Feed the nestlings.

Breathe out. Feed the nestlings.

Breathe in. Feed the nestlings.

Breathe out. Feed the nestlings.

Oh, is there more to life?

It was pouring rain, the kids were in day camp, and I was leaving to see Dr. Wendy. Strapped to the backseat was the closeable wicker picnic basket; inside, tucked into various nests, were the current nestlings—two house wrens, a tufted titmouse, and a chipping sparrow. Riding in carriers in the back of the Jeep were a herring gull and a Canada goose.

"A goose!" I'd said to the man who called. "Can't you find any injured flamingos? I only take healthy adult songbirds, but I'd make an exception for a flamingo."

"What?" he said.

John had not been as easily sidetracked. Appearing just as I was about to back out of the garage, he cast a suspicious look into the car.

"What have you got in there?" he asked.

"Just the nestlings!" I said.

"I don't mean in the basket—I mean in those carriers. Those carriers that seem to be too large for nestlings, unless they're nestling pterodactyls."

"I promise you," I said solemnly. "I will never accept a pterodactyl from a member of the public."

John opened the back door and peered behind the towels covering the carriers. "You're on a slippery slope, aren't you?" he said.

"I'm not keeping the goose," I explained. "I'm taking them both down to Wendy, and she's going to give the goose to another rehabber. But I have to go, because she's waiting for me. Bye!"

I drove slowly, peering through the rain and following a line of traffic. When I was halfway there, the driver two cars ahead of me hit a large woodchuck. He tapped his brakes briefly, then kept going. The driver directly ahead of me made a wide circle around the woodchuck, who stood, fell over, rose drunkenly, and fell again. I yanked the steering wheel to the right and pulled off the road, watching incredulously as car after car avoided the staggering creature and continued on its way, barely slowing down.

Half of my brain ordered me to pull back onto the road and follow their example. The other half shouted, "Fer Chrissakes, *hurry up before some idiot kills him!*"

Cursing, I jumped out of the car, opened the back, and grabbed the blanket covering the gull's crate, also snatching two small cardboard trays that happened to be propped up between the crates. I marched into the middle of the road and threw the blanket over the woodchuck, suddenly realizing that I knew very little about them. I thought furiously, trying to remember the woodchuck questions from my New York State Wildlife Rehabilitation Exam.

Which of the following animals normally hibernates during the winter?

> a) opossum
> b) red squirrel
> c) woodchuck
> d) raccoon

My grandfather once told me that all knowledge has value, although knowing that woodchucks hibernate in the winter did not seem especially valuable to me as I was standing in the middle of the road. Assuming that if I suddenly grabbed a wild woodchuck he would bite me whether or not he had recently been clobbered by a large car, I swaddled him in the blanket, put one cardboard tray under him and one over him, picked him up like an overstuffed sandwich, and carried him back to the Jeep. Cramming the whole package between the two crates, I slammed the back door shut, going on faith that during the next five minutes the woodchuck wouldn't somehow wiggle out of the blanket and start running laps around the car.

During the short drive to the veterinarian's office I considered my situation. As John had noticed, I wasn't doing a very good job at drawing the line. But there was a terrible shortage of bird rehabbers in my area, and I had never been good at saying no to an animal in distress. It's just a facet of my personality, I decided. Some people are unable to pass a chocolate truffle lying on a table without grabbing it and stuffing it into their mouth, and I am unable to pass a woodchuck convulsing on the road without grabbing it and stuffing it into the back of my car. At that particular moment I reached the office, thus concluding my haphazard psychological self-assessment.

I parked the car, unstrapped the basket, and sprinted across the parking lot, trying to avoid jostling the baby birds as I ran. Reaching the door, I flung myself inside, breathless, disheveled, and dripping wet. Janet looked up from her desk, pursing her lips in a desperate attempt not to laugh.

"Lovely day for a picnic," she said.

Wendy, ever positive, walked into the waiting room and regarded me with a look of pleasant surprise, as if I had just strolled in wearing tennis whites and holding a mint julep.

"It's not enough that I have a goose and a gull and all these babies," I told her. "Now I have a *woodchuck*."

"Room number one," said Wendy, without missing a beat. "Need help bringing him in?"

A minute later I hurried back in through the front door, balancing the woodchuck sandwich between two soggy cardboard trays.

"Hey!" caroled Janet as I disappeared into the examination room. "You want some mayo on that groundhog?"

Wendy closed the door and pulled at the blanket, searching for the beast within. An opening appeared and the woodchuck burst into view, blood dripping from his nose, chattering in what I would term an aggressive way even though at the time I had no basis for comparison. Wendy calmly grabbed him by the scruff of the neck, felt around for broken bones, shone a penlight into his eyes, and finally announced that he looked fine but could probably use a shot of cortisone. Depositing the woodchuck gently on the floor in the corner, she tossed the blanket back over him and left the room.

I opened the picnic basket and fed the nestlings, keeping a wary eye on the blanket in the corner of the room. I'd fed the last one and was closing the lid when the blanket started moving ominously. Should I act casual, I wondered, or run back into the waiting room and hide behind Janet? Get ahold of yourself, I told myself sternly. Wildlife is wildlife; it's just this one is huge, hairy, and has giant rodent fangs instead of a beak.

Wendy returned with a hypodermic and a large cardboard box. As soon as she bent down and gave the injection, the woodchuck shot out from under the blanket and ran straight toward me. Leaping into the air, I did a lively Mexican hat dance, silently vowing that someday I would rewrite the New York State Wildlife Rehabilitation Exam with more pertinent questions:

A woodchuck has been hit by a car, dragged into a veterinarian's office, held by the scruff of the neck, given a shot, and is currently racing around the office floor sounding like an enraged lawn mower. Should the rescuer fear that the woodchuck will bite her foot?
    a) yes
    b) no
    c) not if the rescuer has explained to the woodchuck that *she is a bird rehabilitator and doesn't do mammals.*

Wendy intercepted the woodchuck, deposited him into the large cardboard box, and covered the box with my blanket. "There we go!" she said cheerfully. "Now—who's next?"

The goose and the gull both had broken wings. The rehabber Wendy had contacted was a kind woman named Marylyn Eichenholtz, a mammal rehabber who had agreed to provide temporary care for the goose until she could pass her along to a waterbird rehabber up north. That left me with the herring gull, normally a big strapping creature with wide yellow eyes and a bad attitude. This one was thin and weak, having been grounded in a parking lot for two days before someone rescued him. This did not prevent him from methodically biting my hand at every opportunity; it just meant that, at this point, his bite didn't hurt.

"That'll change," promised Wendy. "Believe me, you'll know when he's starting to feel better."

After both birds had been treated and their wings wrapped, Wendy lifted the blanket and regarded the woodchuck.

"I would bring him back where you found him and let him go," she said. "Unless you're really dying to take him home."

"No," I said. "Believe it or not, I'm drawing the line."

I made my way through the waiting room, carrying the gull's crate and the basket of nestlings, while Wendy followed behind me with the woodchuck. "Suzie!" cried Robin Sista, the other receptionist. "Janet tells me you're doing mammals! Can you take a coyote?"

I drove back to where I had found the woodchuck, pulled off the road, and squinted through the downpour at a heavily wooded steep slope. This couldn't have happened next to a nice level field on a sunny day, I thought grumpily. I opened the back of the Jeep, picked up the woodchuck's box, trotted across the road, and began staggering up the slope, trying to maintain enough momentum to keep me moving upward but not so much that I would lose my balance and somersault, box and all, down onto the road—where, no doubt, cars would make a wide circle around me and continue on their way. I finally slowed to a stop and, gasping, set the box down in the mud. Through the drone of the rain I could hear the sound of angry chattering.

I pointed the box uphill, tipped it over, and pulled the blanket away. The woodchuck bolted out, took a few strides, then made a U-turn and started galloping back down toward the road.

"%$@$&*!" I shouted. "You %@$#(# &%*#% &%*$#@!"

I raced after him until the ground gave way and I found myself sliding. The woodchuck looked to the side and found that the human he had spent most of the afternoon trying to get away from was mudsurfing beside him, wobbling dangerously and howling like a beagle. He slammed on the brakes just as I hit an exposed root; eventually I rolled to a stop just short of the road. I looked up groggily and watched as the woodchuck loped up the hill, paused for a brief moment, and disappeared into the woods.

Chapter 15

CLAWS

I looked into the cardboard box, where a beautiful young black-billed cuckoo lay bloodied and gasping for breath.

Once again the bird had ended up with me by default. The finder lived less than ten minutes away and had been given my number by the local Audubon sanctuary.

"Wait a minute," I had interrupted. "How did they know I was doing this? I didn't tell them."

"Got me," the man replied. "Somebody told them."

I gave him three other numbers, but it was a Sunday afternoon and he had been unable to reach anyone.

"Look," I said when he called me back. "The bird needs a certain kind of antibiotic, and I don't have it."

"Can you get it?" he asked. "If I can bring the bird to you, I'll drive to wherever you want and pick up the medicine."

It wasn't that easy. The bird needed a wide-spectrum antibiotic sold through veterinary supply companies, not through regular drugstores. No local veterinary offices were open on Sunday. The closest twenty-four-hour animal emergency groups were at least an hour away, and they weren't going to hand out drugs to a total stranger without a call from a licensed veterinarian. I called the four numbers I had given the finder, leaving the same desperate plea on each machine: "Please call me the second you get this message!"

The doorbell rang, and I opened it to find a middle-aged man and his teenaged son. I invited them in, told them to wait, carried the box into the bathroom, and shut the door; that way if the bird suddenly flew out when I opened the box, he couldn't go far. I pulled back the lid and winced, then felt a hot surge of anger. Although I was relatively new to songbird rehabilitation, I was no stranger to the one-sided war between birds and outdoor cats.

Of all the ways human beings casually slaughter "protected" wildlife, letting domesticated cats outside is by far the most egregious. And the most easily shrugged off. People who wouldn't dream of taking a shotgun and blasting a bird out of a tree let their cats outside, which accomplishes the exact same thing but in a slower and more horrifying way. "Don't scream at these people and call them names," I remembered a rehabber friend telling me. "There's a chance for education here."

I walked deliberately back into the living room. "Why do you let your cat go outside?" I asked.

"Cats," he said. "We have six."

"And they all go outside?" I said, more loudly than I meant to.

"We can't keep them inside," the man replied. "They'd tear up the screens."

"So instead you let them go outside, where they tear up the birds."

"Well," he said philosophically. "It's all part of nature."

"Really!" I said. "Do you feed your cats?"

"Of course."

"Then how is that a part of nature? If a wild animal doesn't catch his own food, he starves to death. If a housecat doesn't catch anything, it goes home and eats Little Friskies. Most cats don't even eat what they kill."

"True, but . . ."

"If a car hits your cat and it drags itself home, do you let nature take its course or do you take it to the vet?"

"She's got you there, Dad," said the son.

"The cats don't always hurt the birds," said the man. "Sometimes they just play with them and let them go."

"Oh, my God," I said, fighting the urge to lure the man into the driveway and run him over with my Jeep. "Cats' mouths and claws are crawling with bacteria. All a cat needs to do is put one tiny little nick into a bird and it's a dead bird, only it'll take her two days to die. If it's springtime, she'll die and all the babies who are waiting for her in their nest will starve to death."

"That bird we just gave you?" said the son. "One of our cats caught it and brought it into the kitchen. One cat had it and the others were trying to grab it away from him."

After giving his son a quick look of irritation, the man unwillingly looked back at me, like a child standing in front of a principal.

"Can you save the bird?" he asked.

"I don't know," I said. "I don't know how badly she's been hurt. I'm going to let her rest before I examine her, and meanwhile I'll try to find some antibiotics."

While the man and his son waited in the kitchen, I tried my last resort.

"Wendy?" I said into the phone. "I'm so sorry to call you at home, but . . ."

"Don't be silly!" she said. "That's why I gave you my number! What's the trouble?"

A few minutes later I returned to the kitchen and handed the man an address.

"Do you know where that road is?" I asked. "About a mile from the turnoff you'll see a yellow mailbox on the right. The antibiotics will be in there."

"Great," said the man. "We'll be right back."

At the door, the son hesitated. "Maybe I could talk to my mom," he said. "But she doesn't care about birds."

I smiled at him. "It's worth a try," I said.

Songbird populations are plummeting. The hardy, opportunistic species that adapt well to human interference are holding their own, but the others are not so lucky. The two greatest threats to songbirds are habitat destruction and outdoor cats. According to the American Bird Conservancy (www.abcbirds .org), there are 90 million pet cats in the United States, and according to one poll, only 35 percent of them are kept exclusively indoors, even though 65 percent of people polled believe that keeping cats indoors is safer and healthier for the cat. During an eighteen-month period, a single cat roaming a wildlife experiment station killed over 1,600 birds and small mammals. A study in England showed that cats wearing bells killed more birds than cats without them; during a study in Kansas, a free-roaming declawed cat killed more birds than the cats with claws. And in addition to all the pet cats out there killing wildlife, there are between 60 and 100 million feral cats, which can have up to three litters of kittens per year. A study in a newsletter published by the California Academy of Sciences concluded that the combined population of outdoor cats kills more than *3 billion birds per year*, and the study was conducted *over ten years ago.*

Those who profess to love the cats they let outside ignore the fact that the average life span of an indoor cat is fifteen to nineteen years, while the life span of cats allowed outside is two to three years. Outdoor cats fall prey to cars, animal attacks (including dogs, wildlife, and other cats), human abuse, poisoning, traps, and a host of diseases, including rabies. Those who "love" their cats might want to show it by keeping them inside, where they are safe and secure. And perhaps those who profess to "love" nature shouldn't advertise their hypocrisy by allowing their pets to slaughter the dwindling wildlife populations around them.

I returned to the cardboard box and quietly opened the lid. The young cuckoo lay on her side, still and lifeless.

When the doorbell rang I took a deep breath, trying to control my rage and frustration. I opened the door, where the man and his son waited once again. "Here are the antibiotics," said the man, handing me the box.

"Come in," I said. "Can you wait a minute?"

I left them for a moment and returned holding the cuckoo. Cradling her gently, I lifted one wing to reveal a deep bloody gash in the delicate pearl gray of her side; following the line of her wing, I showed them where the bones were broken. I parted the soft brown feathers on her back, exposing two deep punctures, and on her leg, where skin and tendons were torn and mangled. I looked up; father and son were staring fixedly at the bird.

"Imagine if you were holding one of your cats the way I'm holding this bird," I said, "and it was your neighbor's dog who did it. What would you do?"

"I'd sue them," blurted the man.

The boy pulled his eyes away from the cuckoo and looked at his father.

"Getting those antibiotics was a nice gesture," I said. "But if you're going to keep letting your cats out, it was meaningless and this bird died for nothing. If you're trying to teach your son responsibility, then keep your cats inside. It will be difficult at first, but they'll learn to deal with it."

"Thank you," the man mumbled, and turned to leave. I stood on the front steps as they got into their car. The boy closed his door and slumped against it, his face turned away. I watched as they drove slowly down the driveway, my heavy hopes for the future of a small group of songbirds all resting on the slender shoulders of a teenaged boy.

Chapter 16

SONGS OF JOY

The herring gull refused to eat.

I offered him canned cat food. I opened his beak, put in a wad of tuna, and closed his beak; he spat it out. I offered him two small freshly caught fish from our pond. I opened his beak, pushed the fish down his throat, then held him gently to give it time to settle. As soon as I let him go, he gave me a disgusted look and threw the fish up onto the floor. I sighed. He bit my hand.

The gull was housed in a medium-size crate in the extra bathroom. A towel covered the back of it, cutting off his view of Daisy and her mallard companion, the blue jays, and, housed in a separate reptarium, two Carolina wren fledglings. Each of them had provoked the gull's interest.

"Forget it," I told him. "Live birds are not on your menu."

Leaving him with an assortment of food items, I took a break so the kids, the nestling songbirds, and I could go shopping for new sneakers. If not for the obsessive records I kept in my red three-ring binder, I would have lost track of who I was lugging around in the wicker picnic basket; for although I kept insisting—more and more weakly—that I didn't do nestling songbirds, they kept coming in. I fed the babies, we drove to the shoe store; Skye fed the babies, we bought the shoes; Mac fed the babies, and we headed off to Burger King. This was when Burger King was still a culinary Mecca, before Mac downgraded it to a biannual event and Skye deemed it the most vile place on earth. We sat in the car line, planned our meals, and watched as the gulls swirled above us.

"I think you should get the herring gull a Whopper with cheese," said Mac.

I stopped. The gull had come from a Cortlandt parking lot, and we were in a Cortlandt parking lot surrounded by gulls. The lightbulb clicked on.

"Get him some fries, too," added Skye.

When we returned home I stood in front of the gull's crate while the kids watched from the doorway. Pulling a french fry from the Burger King bag, I pushed it through the metal grate on the front of the crate, where it fell on top of the small mound of untouched food. Like a striking snake, the gull snatched up the fry, then stood gazing at us expectantly.

"Can I give him another one?" asked Skye, jumping up and down.

"Me, too!" said Mac. "See? Everybody likes Burger King."

Later I ranted to John. "A gull won't eat fresh fish but he'll eat a greasy pile of chemicals! How much more can humans screw up the planet?"

"You should be grateful for that greasy pile of chemicals," said John. "The gull and your children certainly are."

Actually I *was* grateful, for the fast food jump-started the gull's appetite. He'd dig through the nutritious food in search of the lone french fry or the chunk of cheeseburger, and after scarfing it down, would resignedly move on to the defrosted frozen smelt and the vitamin-covered Nine Lives Ocean Whitefish Dinner. As Wendy predicted, I knew he was feeling better by his rapid increase in jaw power.

"Ouch!" I'd grimace, and leave the bathroom rubbing my hand. As it turned out, it was a gesture known well among rehabbers.

"What's the matter with your hand?" asked Joanne, getting out of her car. "Got a gull?"

Joanne was bringing me a fledgling gray catbird. "Cute little guy," she said, "but his feathers are lousy. Woman kept him in a wire cage. Fed him bugs, though, so he should be okay."

Wild birds cannot be released unless they are in perfect feather. Missing, frayed, or broken feathers mean the bird will not fly well enough to avoid predators or, in the case of raptors, catch their prey. Stress bars—the weak, light-colored areas that occur when the bird is sick or starving—will disappear if the bird regains his health and grows a new set of feathers, but until then they preclude the bird from being released. Waterproofing, which prevents the bird from becoming soaked and earthbound during bad weather, can be done only by the bird itself, but must be verified by the rehabilitator before release. Few things are more important to a bird than its feathers, and the little catbird's were a mess.

But except for the caging he had been well fed and cared for, which meant that soon he would go through a molt and grow a fine set of new feathers. With a bit of luck he could be released in September. I let him go in the flight

cage with the robins, finch, waxwing, and sparrow, learning quickly why many rehabbers have such a soft spot for catbirds.

Catbirds, along with mockingbirds and thrashers, actually belong to the Mimid family, known for their gift of mimicry. The repertoire of adult Mimids can include their own songs, calls of other birds, various local sounds, and in the case of catbirds, the hoarse mewing that gives them their name. In personality, catbirds are a bit like Corvids (crows and jays) but without the attitude. Both are incredibly busy. But while crows always seem determined to put one over on you, catbirds simply want to know what you're up to. The little fledgling followed me around the flight cage like a miniature Margaret Mead, studying my every move and occasionally peering quizzically up into my face. If I dropped a piece of string, he'd wait until I was safely out of range, then rush over, grab it in the middle, and twirl it around like a gymnast with a ribbon; when he tired of this routine he'd drag it laboriously up a tree to hook it on a high branch, where he could retrieve it later. He wasn't tame enough to land on my shoulder, but neither was he wild enough to cover up his naturally nosy behavior. Each morning he'd wait until I was busy cleaning the water dishes, then he'd quickly fly over and poke through my food carryall to see what I'd brought for breakfast.

"I don't know what it is," I sighed to the kids. "I just love that little Romeo."

"Romeo!" gasped Mac. "You named the catbird Romeo?"

"Awwwwwww!" said Skye happily. "Then let's keep him!"

"We can't," I said. "He's a wild bird. It wouldn't be fair to keep him in a cage."

"Awwwwwww," the kids chorused, this time in a minor key.

As with the grackles I tried not to play with him, hoping he'd begin to look to other birds for companionship, but occasionally I'd toss a leaf or a pebble and he'd race over to investigate; if the object landed in the water dish, it would provoke a furious bout of bathing. Sometimes the kids and I would loiter outside the flight cage, watching as the catbird studied the behavior of his companions.

One day we watched as the house finch hopped up a large angled branch. Right behind him was Romeo, following intently but keeping a polite two hops distance. When the finch stopped, Romeo stopped; when the finch hopped on, so did Romeo. Finally the finch's patience ran out; he turned around and let out a loud "*a—a—a—a—a—a!*" Romeo, abashed, flew off.

A few days after Romeo arrived, Daisy's little mallard friend took a turn for the worse. In a prescient display of emotional self-protection the kids hadn't given him a name, saying they couldn't decide what to call him. For almost a week we had seen daily improvement. His falls had become less frequent, and he could steady himself by his food dish and eat by himself. He loved the water, where he could stay upright, change direction, and even swim beneath the surface. But that morning he suddenly arched backward, fell heavily, and was unable to push himself back to his feet. He'd fall to the side and spin; when helped to his feet he'd fall again as Daisy, subdued, watched from a corner. His falls became progressively more violent until at last I searched through my small box of wildlife medication and sedated him, cushioning him in my hands until the spasms that wracked his small body subsided. Maggie picked him up at the end of the day for what would be his final trip to the vet, and Daisy was alone once again.

"Will the mallard come back?" asked Skye after Maggie had driven away.

"I don't know, honey," I said. "If the vet can help him, he'll be back."

The subject didn't come up again.

I put another message on my electronic mailing list: "Desperately seeking wood duck(s)." Meanwhile Daisy became more self-sufficient, drowsing in her enclosure when she was by herself, springing into action when the kids tossed her mealworms, paddling around a rubber tub filled with duckweed that I gathered from a local pond. When we put her in the bathtub she'd dive beneath the surface, then rocket around the bottom at astounding speed, looking more like a seal than a duck, and I'd feel a twinge of regret for all the captive ducks with no access to their natural element.

The call came on a sunny Saturday morning a week later. I was planting pachysandra in the protected area outside the front door while Daisy busily hunted for bugs, weaving in and out of my knees and slowing my progress considerably. Skye sat on the front step, reading spooky stories aloud, while Mac rode his bike up and down the driveway. Skye ran for the cordless phone and handed it to me.

"Suzie?" said a pleasant female voice. "My name is Hope Brynes. A friend of mine read your post on the e-mailing list. I have four baby wood ducks, just about your duck's age, and they're all doing well. If you want, I'd be happy to take yours in."

I hung up the phone and relayed the message, trying to appear thrilled with the news. "But she can't go," said Skye. "She's too little to leave us! She's not even three weeks old!"

"That's why she has to go," said Mac. "Otherwise she won't know how to be a duck."

Hope lives far north of me, but one of her volunteers met me in the parking lot of an Albany mall, a two-hour drive from my house. In the parking lot I transferred Daisy to her new carrier and busily explained the details of her care, trying to act professional and matter-of-fact. It worked, at least until I gave them all a final cheerful wave and watched them drive away.

When I pulled my car back into the garage John opened the door, came down the stairs, and peered into my miserable face. "Come on," he said. "The kids have something to make you feel better."

Mac and Skye were waiting in the living room with a boom box and a CD. "It's 'Weird Al' Yankovic!" said Mac, holding up a CD cover showing the fuzzy-haired song satirist staring maniacally into the camera. We all sat down to the unmistakable starting chords of Huey Lewis and The News' 1980s hit, "I Want A New Drug." However, in the capable hands of "Weird Al," the song had been transformed to "I Want a New Duck."

I want a new duck
One that won't try to bite

One that won't chew a hole in my socks
One that won't quack all night

We laughed until our sides ached, playing it over and over again. All we can do is offer these unlucky wild ones a second chance and then let them go, no matter where they're headed. Some are like Daisy, who would remain friendly to her human caretakers but bond with her new siblings so strongly that in the fall, they'd all take flight and disappear together. Others are like Daisy's little companion, whose eventual freedom meant liberation from a body that couldn't be fixed, despite our best efforts. As we all lay giggling on the living room floor I briefly wondered if the joy that Mac and Skye might find in the freedom of a newly released wild bird could ever make up for the sense of loss they'd feel as it flew away. And if the life I was giving them, with its constant themes of life and death, held too much sadness. At that moment Skye turned the volume up and she and Mac dissolved into fresh gales of laughter.

One that won't drive me crazy
Waddling all around
One who'll teach me how to swim
And help me not to drown

Perhaps, I thought, you just have to take your joy when it appears, even if it comes disguised as a "Weird Al" song.

Chapter 17

GRADUATION DAY

"Ruth!" I said on the phone to my friend, the former rock star. "Remember that robin you brought me? He's all better. I'm going to release him."

"Ah, that's terrific!" she said. "Where? You're not taking him back to that same field, are you? He'll just get his butt kicked all over again and then I'll find him and have to bring him back to you."

"Maybe not," I said. "Maybe he's had time to rethink his strategy. But I'll let him go here. If he really wants to go back for another round it's only a mile or so away. He has a little entourage of young robins, so if I release them all together maybe he'll decide to stick around."

"That's cool," said Ruth. "Go get him and put him on the phone and I'll sing 'Free Bird.' You know, for inspiration."

The kitchen's bay window looks out over a small strip of "lawn"—our generous term for a motley collection of weeds, moss, and the occasional struggling patch of actual grass—and on to a wide slope that rises into the woods. Dotted with huge, immovable rocks and planted with berry bushes and bird-friendly groundcover, the slope sports two bird feeders and a suet holder and is a busy place. From the kitchen table we can watch the interactions of black-capped chickadees and tufted titmice; downy and red-bellied woodpeckers; white-throated, song, and fox sparrows; white- and red-breasted nuthatches; mourning doves; Carolina and house wrens; and brown creepers. One day a

Cooper's hawk will storm into the yard, sending the songbirds diving for cover; a few days later we'll see the hawk race away into the woods, furiously pursued by a pack of crows.

The eastern phoebes return from their winter vacation on April 1, announcing their arrival from the shadbush just beginning to bud. By the end of April we're waiting for the rose-breasted grosbeaks, who arrive two days before the ruby-throated hummingbirds and are occasionally trailed by an indigo bunting or two. By May we struggle to keep the feeders filled for the various migrants who stop, fuel up for a few days, and continue on their way north. By the time the mountain laurel bursts into bloom the slate-colored juncos have long since headed north, only to return after most of the summer birds have followed the warm weather south. But the grand spectacle comes at the end of October, when one of us will hear a rattling, grating waterfall of sound that seems to descend and envelop the house; we all rush out to the deck to find the woods black with common grackles—hundreds and hundreds of them, their iridescent feathers shimmering in the light, their harsh voices filling the autumn air. When they see us they startle and lift off as one, the sound of their wings like a spinnaker filled with a sudden gust of wind.

Into this avian merry-go-round I wanted to introduce rehabilitated songbirds. My yard was a good place for soft releases: filled with other songbirds, access to food and cover, surrounded by woodland but with a sunny field a few hundred yards away. The kids and I built another bug pit near one of the feeders, and we were ready to go.

We released the robins early one sunny afternoon, filling the bug pit with earthworms and mealworms and then setting the crates nearby. "This is it!" I said to the kids.

"You're not going to cry, are you?" said Mac.

"Why would she cry?" asked Skye.

"Remember all those hawk releases we used to go to? All the grown-ups were crying."

"Sometimes they're happy tears," I said. "People who take care of birds work so hard, and when everything turns out right they get all emotional. But even the people who haven't taken care of the birds sometimes get weepy because letting a bird go is symbolic, and people see themselves in the bird. Maybe something hurt you and knocked you down, and in spite of the odds you've gotten back up and are trying again. Maybe you're letting go of one part of your life and starting another. Maybe you're trying to let your hopes and dreams take flight."

The kids stared at me. I wanted them to see how much more there was to bird rehabilitation than the day-to-day care, to appreciate why people put themselves through so much just so they could watch a healed bird fly away. I wanted them to know that when I walked into the flight cage I felt as if I were entering Skye's world, as if her kelpie had transformed my man-made building into a land filled with fairies and surrounded by magic, a lush green habitat home to the breathtaking beauty of a robin and the ephemeral trill of a waxwing. And I wanted them to know that my granted wish, like that of a fairy princess, was only temporary; before long, all those I had wished for would take to the sky and disappear.

Mac stared off into the distance; Skye gave me a searching look and took a deep breath.

"Can I have macaroni and cheese for dinner?" she asked.

When we opened the doors the adult robin rocketed up into a nearby hemlock; two of the fledglings followed him, another took cover under a juniper bush, and the last one surveyed the scene from a mountain laurel. I wondered if the young ones remembered freedom, having come into captivity as fledglings only several weeks earlier. I crouched by the bug pit, sifting worms through my fingers and demonstrating where the fast food was located, feeling as if I should mark the occasion by singing "Free Bird"—as Ruth had said, for inspiration.

"What are you singing?" asked Skye. "And why are you singing with that accent?"

"It's a Lynyrd Skynyrd song," I said. "It's called . . ."

"Lynyrd Skynyrd!" said Mac. "Is this another one of those old guys from back when you were born?"

"Oh, never mind," I said.

Later we released the song sparrow, who had been left in a box on the doorstep of a local animal hospital and eventually found her way to Joanne, and finally to me. My yard was a good release area for her, as it was frequented by other song sparrows, and I hoped she would find a new group.

I had taken the house finch back to Alan two days before, as the finch wasn't flying well and I was worried that the eye injury might have affected his vision.

"There's no conjunctivitis," Alan had said. "He can see out of that eye, although I can't guarantee how well. His wing is locked up again, though."

"All right." I sighed. "I stopped doing the physical therapy because he was getting too stressed out. I can just catch and release him in the flight—he doesn't mind that as much."

To anyone who didn't know him, Alan's face was pleasantly expressionless. But I could see the infinitesimal lift of the eyebrow, the almost imperceptible tugging at the corner of his mouth.

"Quit it!" I said. "He's my first songbird and he's eating well and I don't think he's in any pain, so I'm not going to give up on him."

"Good for you!" he said emphatically, then, conceding the battle, grinned and shook his head.

The day we released the robins was the day the flight-cage-go-round began. One might assume that unlike raptors, who tend to eat their flightmates if not paired correctly, songbirds belong to one big peaceable kingdom and can happily share a large space. Naturally, this is not the case; that would make the rehabber's life too easy.

Robins, who can be quite aggressive with each other—as ours had learned—are not aggressive with other species, which was why they could live in the same enclosure with the finch, the sparrow, the catbird, and the waxwing. But coming up and needing flight cage space were the six blue jays, two Carolina wrens, and three purple finches (all currently residing in the extra bathroom), and a wood thrush and a downy woodpecker (coming from other rehabbers). Not to forget the grackles, who were behind flight door number one.

The wrens, finches, thrush, and woodpecker could all go in with the house finch, catbird, and waxwing, since all are gentle songbird species. But blue jays, who are in the crow family, are aggressive and can't be housed with songbirds; and neither can grackles, which are actually Icterids. Fox and field sparrows are gentle, but not house sparrows! Hairy and red-bellied woodpeckers are easygoing, but yellow-shafted flickers are killers! Catbirds are kind little birds but are in the same family as mockingbirds, who will beat up anything in sight! The wood thrush is related to the robin, but the robins just left!

"I have an idea," said John. "Put the herring gull out there, and that'll take care of all of them."

Had the jays and grackles been adults, I might have worried about putting them together. But they were all juveniles, they had no experience with other birds, and any territorial feelings the grackles might have about their adopted habitat would be offset by the fact that they were outnumbered by the jays.

The following morning we carried the jays, who were finally eating on their own, out to the flight cage and unzipped their reptarium. When they flew up to various perches the grackles froze, their eyes glued to the intruders, looking like two members of an undiscovered jungle tribe who had somehow stumbled on the cast of Blue Man Group. Although the jays were probably just as taken aback by the sight of the dark, yellow-eyed birds, they had no time for reflection; faced with over two hundred square feet of space they paused for a moment, then turned into kids on the last day of school. They flew sideways, ricocheted off of the sides of the flight, and played chicken with my head. They'd land near one of the grackles, then immediately spring off the perch and giddily flap away, as if they had taunted Death itself and lived to tell about it.

During the day I spotted several song sparrows weaving their way between the bushes on the slope, searching for insects and picking up the occasional seed; and I wished that I were capable of identifying the one I had just released among a crowd of—to my undiscerning human eye—exact duplicates. I had seen all five robins just before dusk the previous evening, perching on various tree limbs, pulling bugs from the lawn and snatching mealworms from the bug pit. Today I had seen all four juveniles but not the adult. John and I sat on the deck at the end of the day holding glasses of wine, watching as the sun descended and the four young robins traversed the slope.

"I think the adult has gone home," I said. "I hope he'll be all right."

"Don't worry," said John. "He'll be fine. He's a wild bird."

I had released many birds during my years as a wildlife volunteer, but I had been part of a group effort; I had not been solely responsible for any of them. I had been positive and matter-of-fact about releasing the songbirds in front of the kids, but now, alone with John, my bravado began to splinter.

"But what about the fledglings?" I said. "Now they don't have anyone to show them the ropes. They're out there all alone. It's dangerous."

"They have each other," John insisted. "They're strong and healthy, and they know how to get food. They'll figure out the rest."

"But what about the song sparrow? It's not her territory. She's in a brand-new place. What happens if . . . ."

"You can't do this," said John. "Once you release them, your job is done."

I watched the birds, feeling my neck knotting in silent protest. I had given both the adult robin and the sparrow a second chance, but I would never know what they did with it. Would they live long and healthy lives? And what about the fledglings? Would they all reach adulthood? I had done with each of them what a rehabber is supposed to do: I brought them back, then I let them go. But how could I let them go after I'd let them go?

John held up his glass. "Congratulations on your songbird release."

I tapped my glass against his. "Thanks," I said, and hoped for the best.

Chapter 18

WHY WE DO THE THINGS WE DO

As much as I had wanted to return to the goshawk nest, time passed too quickly and by the time I returned all the nestlings were gone. As I ran through the woods I could hear the young ones' wailing cries, and occasionally I'd catch a glimpse of a large dark bird flying swiftly through the trees. I wanted nothing more than to sit under an old hemlock for an entire afternoon and wait for them, but the days of long, lazy afternoons had vanished.

I moved the two fledgling Carolina wrens and three newly arrived fledgling purple finches into the flight cage with the house finch, the catbird, and the waxwing. Maggie arrived with the wood thrush, an adult whose broken wing had healed, and the downy woodpecker, whom she had raised. When I entered the flight cage the wrens would freeze and the finches would quickly cluster around the house finch, who had assumed the role of den mother; the thrush would vanish into the underbrush, not to emerge until I had left; and the woodpecker would hike jauntily along the mesh until she was two or three feet away from me, determine that I wasn't all that interesting, and turn around and go about her business.

Next door things were not so genteel. If the songbird flight was an English drawing room, the jay/grackle flight was the local pub: noisy, boisterous, and filled with outsized personalities looking for trouble. The grackles were slightly older and a bit larger than the jays, and occasionally seemed to be able to intimidate them through sheer size and agility. But the jays never seemed to run out of creative deviltry. They'd steal the toys, hide the food, and, waiting until one of the grackles was distracted, rush up behind them—simply, it seemed, in order to watch them jump. Once I found one of the jays flying about with a grackle tail feather in his beak; although I wanted to assume that the feather had molted naturally, I couldn't guarantee the jay hadn't crept up behind the grackle and yanked it out.

The one member of the group who didn't conform to type was Norbert, the smallest blue jay, who always reminded me of the quiet, bookish mascot of a street gang. Although from time to time I would see him bathing, flying around the flight cage, and interacting with the other birds, he preferred to watch the action from a sunny perch. He had somehow convinced at least two of the other jays to feed him, and he would sit on a wide branch like a prince on a throne and graciously accept their offerings. At one point one of Norbert's feet had suddenly and briefly become swollen, and for a few days I had to catch him for a daily foot check. This was easier said than done; for all his sedentary ways, he was surprisingly quick when he wanted to be. He dodged my net, wiggled out of my hands, and made such a commotion that by the time I caught him everyone else was in an uproar. One day he flew to a corner, and as I approached him all five of the other jays flew over and perched on branches between us. They stared balefully at me, silent and motionless, like a small squadron of bodyguards protecting their client from a familiar and not very threatening enemy.

When it came to downtime the two species would separate. The jays gathered into a small flock at one end of the flight, while the grackles perched together at the other. Occasionally on a hot, lazy afternoon I would quietly enter the flight cage and sit on a log in an unoccupied corner, where I could watch them without being obtrusive. The grackles were more aware of me and, being recent allies instead of siblings, not as affectionate with each other. Perching together but only occasionally touching, they would fluff out their feathers and relax, but only rarely did I see them close their eyes.

The blue jays, however, would position themselves comfortably, rustle their feathers, preen each other gently, and, in time, lean against each other and close their eyes; sometimes one would even tuck his head beneath a wing. And Norbert would dream. Drowsily nestling down into a soft circle of feathers, he would point his beak straight up at the ceiling and chortle and coo to himself—eyes closed, fast asleep.

○ ○ ○ ○ ○

"You feed those things every half an hour?" asked the woman incredulously, staring down into the wicker picnic basket. "Are you out of your mind?"

I was in my usual Saturday morning position, sitting in a collapsible chair under a tree and watching the kids play soccer, occasionally glancing at my watch and doling out mealworms to the waiting nestlings. Every now and then a parent would stop by and peer into the basket. When they likened me to Mother Teresa, I'd envision myself surrounded by golden rays of light, wearing a halo and smiling beatifically; when they questioned my sanity, I'd envision myself in an old-fashioned black-and-white striped prison uniform, scowling ferociously, holding a ball and chain instead of a wicker basket.

What was the matter with me, anyway?

I was so busy that I rarely had time for reflection, but during the occasional lull—or when asked a pointed question—I would try to consider the Big Picture. What good was I doing, exactly? Would any of the nestlings I labored over that summer survive to adulthood, then go on to breed? Could they ever overcome the huge hurdle of being raised and released by an alien species? Jayne bands her fledgling birds before releasing them and sees many of them year after year, so I know if I raise them right they have a decent chance. But with my limited time and resources, why was I raising common and plentiful songbirds instead of working to conserve habitat? Or raising money for endangered species? Or fighting global warming?

If we're talking about the greatest return for one's effort, then never mind the baby birds—why rehab wildlife at all?

Exactly, critics say; rehabbers are nothing but a bunch of bunny-huggers wasting their time. Populations are what count, not individuals. It's not worth the effort.

First, when any potential critic looks down on me from his lofty position and deigns to grade my effort, I tend to ignore (or mock) him out of principle. But this is an argument easily won. Although wildlife rehabilitation begins with the individual, there is a ripple effect that extends far beyond the single animal. If critics of wildlife rehabilitation are looking for numbers, they will find them not in the release rates of a single rehabilitator but in the numbers of people who have been reached and educated because of her (or him).

This is not to denigrate the principle of wildlife rehabilitation because, unlike biologists, we rehabbers do believe in the value of the individual. It is easy to dismiss an unfamiliar group, whether it be a flock of bluebirds, a herd of elephants, or a village of Tanzanians. I have never seen a bluebird, one might say, I have never been near an elephant, and I don't know any Tanzanians. But all that changes with contact and familiarity.

My friend India Howell, with whom I lived on the farm in Maine, took a trip to climb Mount Kilimanjaro when she was in her mid-forties. She had never been to Africa and knew little about the Tanzanian people, outside of the occasional news reports of villages devastated by poverty and AIDS. But while she was on vacation she was offered a job as manager of a safari company, and when she moved to Arusha she began to encounter the street kids mentioned in the news reports. Once befriended they were no longer nameless and faceless, and no longer the blurry part of a problem too large to address. India founded and now runs the Rift Valley Children's Village, an orphanage outside Keratu, and she channels certain donations to help the surrounding villages.

Wildlife rehabilitators find themselves in the same position but faced with a more skeptical public, many of whom seem to believe that wild animals are little more than programmed robots. Some loudly and indignantly question why rehabbers "waste" their time with animals when they could be helping people, a query even more absurd than asking a pilot why he or she is not a firefighter. Just as India saw something in the children of Tanzania that she could not turn away from, so rehabilitators see something in a wild animal that can be found nowhere else. We crave a connection—no matter how brief or tenuous—with a wild creature, and we are willing to play by rules that seem designed to break our hearts in order to do it.

We clean, feed, study, attend conferences, amass arcane knowledge, and learn to handle the creatures who fear us. Our triumph is to accept an injured wild animal, treat its injuries, carefully learn each one of its quirks and preferences, help it heal, and then let it go. If things go according to plan, we will never see it again.

Somehow, this is enough.

"Do you ever fall in love with the animals you take care of?" I asked a rehabilitator, naively, years and years ago.

She gave me a small, rueful smile. "Every single one," she said.

What rehabilitators learn all too quickly is that each animal, each bird who comes through the door is unique. Species may share general traits, but each individual is different, each one is memorable. And as soon as this becomes clear, the enormity of what humankind is doing to the natural world becomes all the more harrowing.

Critics may look for numbers, but from that point of view all nonprofit work is the veritable drop in the bucket. Millions are under seige; what's the point of helping fifty, or a hundred, or a thousand? The point is in the value of the individual, and in the ensuing ripple effect. The drop in the bucket is the convulsing mockingbird; the ripple effect is that a woman brings it to a rehabilitator, who convinces the woman to stop using pesticides on her lawn, and the woman returns home and convinces her neighbors to do the same. The drop in the bucket is the nest of owlets fallen from a chain-sawed tree; the ripple effect is that a man brings them to a rehabilitator, who dissuades the man from clear-cutting the rest of his property, and the man brings up the effects of clear-cutting at the next town board meeting.

Ninety-five percent of wildlife injuries are the direct result of human activity. Our recent national leaders have championed business and money at the expense of everything else, and deemed a robin's life—since it has "no commercial value"—barely worth noticing. If there is nowhere for a member of the public to bring a single injured wild animal, then the animals' collective lives will become even cheaper than they already are. If the average person's initial concern over an injured bird is met with nothing but shrugs and apathy, he will conclude that wildlife really isn't worth saving, and the war over intrinsic value will truly be lost.

There is a story that every rehabilitator knows, written by the renowned anthropologist, ecologist, and writer Loren Eiseley. A boy walks down a beach covered with stranded starfish, methodically picking them up and throwing them back into the sea. An old man sees him and says, "Why are you wasting your time? There must be thousands of them! How can what you're doing possibly make a difference?"

The boy picks up another starfish, tosses it into the sea, and regards the man. "It made a difference to that one," he says.

I sat in my chair at the soccer game, alternately feeding and cheering, unable to reduce my tweezers and worms and picnic basket filled with hungry nestlings to a clever and convincing sound bite. Eventually I would reach the point of admitting that I was just too exhausted to raise children and songbirds at the same time, but halfway through my first summer I still believed that I had the will and the energy to do anything.

"Are you out of your mind?" asked the woman.

"Yeah," I said with a shrug. "I guess I am."

Chapter 19

SONGS OF INSPIRATION

There's a fine art to begging. One must act suitably pitiful for the easy mark, yet be ready to turn on the aggression for a tougher crowd. And should the need arise, it always helps to be able to pack it in, get your own food, and feed yourself.

One of the most skillful beggars I've ever encountered was Orangina, an orphaned female northern cardinal raised by another rehabber and named for her bright orange beak. She arrived on a cloudy summer morning, slender and wide-eyed, her crest held at a tentative angle. I put her into the flight cage with the congenial and easygoing songbird group. Cardinals, with their thick, seed-cracking beaks, are actually large finches, so at least she had a few distant cousins among the new crowd.

Delighted to have 200 square feet of flying space yet slightly intimidated by her new home, Orangina initially perched quietly on the outskirts of the group. Whenever I'd enter the flight, though, she'd spring into action: fixing me with a melting stare, she'd hunker down on a branch, make heartrending little trilling sounds, and flutter her wings so weakly I was sure she was dying of hunger. Normally I thought of myself as immune to this sort of nonsense—once a fledgling can feed itself, it should do so—but the delicate, tremulous Orangina sliced through my resolve like a steel machete. I'd give her a bite of food, then another, reasoning that she might be having adjustment problems;

then I'd leave the flight, peer back in through the door, and see her happily hopping about, expertly scarfing up seeds and fruit.

Within a day another beggar arrived, but this one wanted nothing to do with me. "I hear you have a waxwing," said Joanne's voice through my telephone. "You want another one?"

Joanne's waxwing had been found as a nestling—perfectly healthy, but minus her legal guardians—on the sidewalk in front of a bar in a large urban area. "It was one of those miracle births," she said. "No nest. Not even a tree. Nothing but concrete, and there's this little cedar waxwing. Amazing she didn't get stepped on." Now almost grown and eating on her own, she just needed a flight cage and another waxwing, both of which I could provide.

Once again I headed out to the flight cage, flanked by Mac and Skye and carrying a cardboard carrier. We passed the Parrot Gazebo, where Zack and Mario were drowsing in the sun. Mario saw us and snapped to attention. "War!" he shouted.

"There he goes again," said Mac. "He must know you have a bird in that box."

The parrots have their own miniature flight cage, an eight-foot cube of wood and chicken wire, built after Zack nearly ended up a meal for a broad-winged hawk. At that point Zack's wings were clipped, and oblivious to the danger, I used to take him outside. One afternoon I was sitting on the deck while Zack, less than two feet away, cruised the railing. What I failed to realize was that the hungry broadwing circling above the house had a clear view of Zack, but thanks to the large canvas umbrella over my head, no view of me. When the hawk dove for him, Zack let out a scream that froze my blood and launched himself off the deck; because of his clipped wings, he quickly lost momentum and coasted to the ground. The hawk expected Zack to fly straight away and overshot his target, giving me time to scramble up the railing, hurl myself off the deck, and shrieking like a lunatic, race over and gather up the traumatized parrot. Whether Zack was more upset by the hawk or by my rescue was anyone's guess, but the next morning a carpenter arrived, and soon my carefully hoarded clothing budget disappeared into the hawk-proof Parrot Gazebo. Filled with natural branches, ropes, and hanging toys, it was nestled partway under a healthy hemlock tree, which provided areas of both sun and shade.

"I have names for the waxwings," I said. "Yin and Yang. The balancing forces of the universe."

"But they're just tiny little birds," protested Skye.

I was eager to see how the two waxwings would interact, as neither had had any contact with its own species since a few days after hatching. While Mac and Skye watched from outside the flight cage, I opened the carrier and the new waxwing flew up onto a branch. Clearly taken aback, Yin gazed at the clustered group of finches and the two wrens, at the thrush, the woodpecker, the catbird, and the cardinal. Finally she turned her head, saw the other waxwing, and like a miniature homing device, flew to his side.

It was obvious that Yin had found her parent, even though Yang was only a week or so older than she was. Beak open, wings vibrating so violently I thought she'd fall off the branch, Yin was the frenzied fledgling that few adult birds can resist. But Yang, being a fledgling himself, managed to resist her. Each time Yin landed beside him, begging piteously, Yang stiffened into a what-fresh-hell-is-this pose of personal affront and flew off to the other end of the flight. Although perfectly capable of feeding herself, Yin followed doggedly, grimly determined to be adopted. I wondered if Orangina would take notice and transfer her begging behavior to the beleaguered Yang, but she remained true to me. At least, for the time being. For the next ten minutes I stayed in the flight cage while Yin pursued Yang, Orangina pursued me, and the rest of the birds watched with interest.

"I think the big one's going to get tired of running away from the little one," said Mac.

"That's what I'm counting on," I replied.

On the way back to the house we paused at the Parrot Gazebo. "Shower time!" I cried. Mac ran to put on the stereo while Skye unrolled the hose. Our parrot shower history is a tangled tale of emergent souls, of song and redemption, all ultimately resolved with Aretha Franklin's version of "Chain of Fools."

Zack came to live with us when he was three months old. Mario, however, was a rescue: he arrived when he was five years old and we were his fourth home. My friend Carol Speier had raised him from a chick, sold him to what she thought was a good home, then found him years later sitting alone and dejected in the back of a feed store. She identified him by a toe on his left foot, partially missing since birth. She took him home, called me with his sad story, and I smuggled him into the house when John was on a trip.

"When Daddy gets home, don't say anything about the new parrot," I had said to the kids, who were then three and four. "Just give him some time to relax, and we'll surprise him later."

"Okay!" the kids agreed enthusiastically. Minutes later John walked into the house to a giddy chorus of *"Wait till you see our new bird!"*

At first, there could not have been more of a difference between the two parrots. Zack swaggered through the house like a feathered James Cagney, constantly demanding all kinds of attention and immediately screaming and biting us if he didn't get it. Mario, on the other hand, sat quietly on his perch, watchful and reserved, accepting attention if it was offered but being careful not to ask for it. If Zack was the house id, Mario was its superego.

Then I put on a Motown tape.

I have a vast collection of old cassette tapes, each one a medley of favorite songs I recorded from albums that bit the dust years ago. I'm sure some techno-whiz could transfer them all onto CDs or into an iPod, or through some sort of singing telephone, but the whole idea just gives me a headache; instead, I guard my cassette collection, dreading the day when one will break and I will be forced to try to duplicate it.

When Mario joined the household there was almost always music playing, the type depending on the time and mood of the participants. Skye was famous for her frighteningly dead-on imitation of Joe Cocker singing his gravelly chorus of "You Can Leave Your Hat On," although I had to keep hiding the tape so she wouldn't discover the rest of the lyrics. John went for 1960s rock, featuring Jimi Hendrix's crashing guitar solos or, if he was feeling moody and existential, the Doors' "The End." Zack was partial to punk-funk and synthesizer bands; blasting Talking Heads' "Swamp" or Soft Cell's "Tainted Love" would send him flying through the house, eyes flashing, screaming enthusiastically. My own musical favorites could fill an entire bookcase, but whenever I was feeling tragic and put-upon, I would return to Bruce Springsteen's "New York City Serenade," which I would play thirty or forty times in a row after donning headphones. The only family member who didn't seem to care what was playing was Mac, who would listen good-naturedly to anything.

Whenever I put on some particularly raucous rock 'n' roll, Mario would stare at me while I drummed on the table and sang along, looking like a nerdy neighborhood kid who wants to join the fun but has no idea how to do it. Finally one morning, when the kids were at school and I was faced with the onerous job of cleaning the kitchen, I slid my favorite Motown tape into the deck. The first song is The Temptations' "Ain't Too Proud to Beg," a song which elicits from me a Pavlovian response: as soon as David Ruffin shouts that first line of heartbreak, I'm ready to party.

As it turned out, so was Mario. Dancing around the kitchen with Zack on my shoulder, I turned to find Mario bobbing his head in time to the music. The louder I sang, the more energetic he became; he seemed particularly entranced by my rendition of "I Can't Get Next to You," even though I couldn't hit Eddie Kendricks's high notes if my life depended on it.

Belting out the soul-filled "I've Got to Use My Imagination" along with Gladys Knight, I raised my arms and mimed the agony of good love gone bad, inspiring Mario to swing upside down from a rope toy he'd previously ignored. During "Son of a Preacher Man" he climbed back up to his perch and whistled furiously, even though Dusty Springfield—being a white girl from Britain—had no business being on a Motown tape at all. Like an astonished 1950s-era parent who watched Elvis transform her quiet child, I watched as Percy Sledge brought out the party animal in Mario.

This was when I got carried away.

Hearing the sound effects of "I Wish It Would Rain," I had a brilliant idea and scurried off to get the plant mister. I squirted it energetically above Mario's head, covering him with a fine shower of droplets. Suddenly I stopped dead and held my breath, appalled at what I had done.

African greys are notorious for their opinions about water. Although parrots raised by people who know what they're doing almost always love a bath, a bad experience can frighten them and drastically change their opinion. Greys, being especially sensitive and emotional, tend to go overboard; either they love to bathe or they act as if you're trying to douse them with lighter fluid. At that point, I had no idea if Mario liked water or not. I waited, wondering if the party was about to come to a grinding halt.

For what seemed like a long time Mario stood still. Then he rustled his wings, shook his tail feathers, and started bobbing his head to "Dock of the Bay."

The party was still on.

Ten minutes later I had dragged one of those long plastic clothing storage units out from under my bed, dumped all the clothes out onto the floor, filled it with two or three inches of water, and lugged it over to the kitchen table, where Mario watched the gleefully screaming Zack perch on its edge, jump in, flap his wings until the water was flying, and jump back out again. After Zack had enough, I carried Mario over. Stepping uncertainly onto the edge, he hesitated.

And at that exact moment "Chain of Fools" began. Aretha Franklin's soaring voice cut through the brief silence, and Mario jumped into the pool. Not only did my kitchen remain uncleaned, it ended up with an inch of water on the floor. But it was worth it; Mario had crossed the bathing divide and added his musical preference to our already cacophonous household.

A hot day in August was even better, with the sun shining, the music coming through the deck speakers, and no kitchen floor to have to clean up. Skye turned on the hose as the first song began.

"She's a very kinky girl!" howled Rick James. "The kind you don't take home to mother!"

"Hey!" shouted Skye. "That's not the Motown tape!"

"I couldn't find it," called Mac.

"No problem," I said. "Technically, it's still Motown. Just twenty-some years later."

Skye twisted the nozzle and aimed the hose into the air, where it fell like rain onto the gyrating parrots. Soon we were all contorting ourselves around the cage, dancing in a way that delighted the birds but probably would have baffled the songwriter. Kids don't listen to lyrics, I assured myself; what's important is the festive moment and the punk-funk beat.

"She's a super freak! Super freak!" caroled Skye.

Chapter 20
SUMMER'S END

"When are you going to get that gull out of the bathroom?" said John. "The whole house is starting to smell like fish."

"I don't have anywhere else to put him," I replied. "His bandage is off and his wing looks good, but both flights are full."

"I thought you were going to release the blue jays and the grackles."

"I am. I will. I am. I'll do it. Really."

"Tomorrow," said John firmly. "They're all ready, it's going to be a beautiful day, and it's not going to rain until the middle of next week. It's a perfect time."

"And then let's go to the movies," said Mac. "We never go to the movies anymore."

"I'm sorry!" I said. "We haven't gone to the movies much this summer. But the last batch of babies is almost eating on their own, and in four or five days I'll be free. Free as a bird!"

"A bird who's not in rehab," said Skye.

Late the following morning I caught all eight birds, put them in three different carriers, and brought them to the slope outside the kitchen window. I tried to remember when I had been as determined to hide my complete, almost paralyzing apprehension; the moment before I had proved my female superiority to my male college friends by jumping off a forty-foot bridge into the local river couldn't even come close. The slope area was busy; two of the released robins were poking through the bug pit. Wearing what I hoped was not an insane grin, I nodded to the kids and together we opened the carriers.

The six jays flew upward, each one a crazy quilt of brilliant blues against the dark green of the spruce trees. The grackles followed, their solemn coloring offset by their bright yellow eyes. For long moments they regarded the sky with what must have been awe, having suddenly and mysteriously become part of a world without boundaries. As they began to explore their new territory we all marveled at their swiftness and celebrated their newfound freedom. My moment was bittersweet, however; all these young birds had been raised in captivity, and by releasing them I could no longer protect them. At least the robins had spent time with an adult, even if the adult had stuck around only for twenty-four hours after I released him. My concern mushroomed; I envisioned all the fledglings sitting at tiny desks, watching me with alarm as I pointed to the drawings on a blackboard. "This one right here is a hawk—watch out! This is an owl—bad news! And this is a raccoon—major problems!"

There were serious gaps in their education. Leaving John and the kids outside, I hurried inside and dialed the phone.

"Jayne!" I said. "I just released four robins, two grackles, and six jays and I'm a nervous wreck. How do you deal with this? What if a hawk comes by?"

"You raptor people!" snorted Jayne. "Welcome to my world, babe! There's nothing you can do except wish them luck. Those damned raptors—every time I release a fledgling songbird it ages me ten years."

"Ten years, huh?" I said. "Then you must be about eighty thousand years old by now."

"I will be by the end of the summer," said Jayne.

I didn't want the kids to worry about the fledglings, so I waited until they were busy on the trampoline before dragging John into the garage.

"Don't tell the kids," I whispered. "But I'm really worried about hawks! What if one catches one of the babies and eats him?"

"What if you get run over by an eighteen-wheeler?" he countered. "Will you stop it? Give these birds some credit! If you're standing there in the yard—I mean you personally—and some huge thing with a curved beak and big talons comes flying at you, are you going to just stand there or are you going to run away?"

"Does he need food?"

"Who?"

"The thing with the curved beak and the big talons! Is he flying at me because he wants me to go get him some food?"

John gave me a baffled look. "Let's stay on topic here," he said. "You're worrying about the songbirds getting eaten at the same time you're worrying about the raptors not getting enough to eat. Something's got to give here, and I think it might be your sanity."

"Or yours," I said.

Realizing that I had transformed the traditional predator-prey relationship into the more problematic predator-prey-rehabber relationship, I wondered what Shakespeare would have said about it and e-mailed Ed. I received a quick reply.

"'O! that way madness lies,'" he wrote, then cited *King Lear* and proceeded to tell me that I had become the fledglings' vestigial organ. "But telling you not to care is like telling the fledglings not to fly," he finished. "So I would say the best thing for you to do is go pour yourself a nice big glass of gin."

At 2:30 that morning I was awakened from a fitful sleep by a horrible noise right outside my window: a cross between a scream, a hiss, and several huge nails being scraped across a blackboard. I sat upright, wondering if I'd dreamed it. I heard it again and shot out of bed, grabbed a flashlight, and bolted out the front door, all before I really knew what I was doing. I knew what the sound was, though: a barred owl.

I ran into the front yard and shone the flashlight up into a grove of trees. Perched calmly about six feet apart were two young barreds, dark-eyed, luxuriously feathered, and breathtakingly beautiful, staring at me with what appeared to be grave concern. One opened his beak and repeated the sound that had awakened me, which was then echoed by his companion. I was light-headed with relief that they were merely conversing and not discussing matters over a fledgling dinner; at the same time I wondered whether they had mastered their hunting skills and if they needed a snack. I decided that what they didn't need was me shining a flashlight into their faces, and I returned to bed.

Our yard became a bird carousel. We ate breakfast on the deck, watching as the various fledglings acclimated to the outside world. We kept the bug pit full and put a plate of food and a pan of water under a low-hanging, protective hemlock branch. The robins weren't afraid of us, but neither were they especially eager for our company; we would see them bustling about, and if we approached too closely they'd fly to a nearby tree. The grackles and jays, however, were more open to interaction. Mac would shout, "Pay attention, you birds! There's a bug down here!" and toss them live crickets to encourage them to forage, then watch in satisfaction as they snatched grasshoppers from the weeds and caterpillars from the trees. The deck was their beach, where Skye once found Harry Potter acting as if he were succumbing to a particularly nasty form of poison. "Look!" she screamed in a panic, "Harry's dying!" She raced over to where the little jay was sitting at an odd angle, beak open, every feather fluffed out, as if caught in some awful rictus. Harry saw her coming, snapped his beak shut, slicked down his feathers, and took off like a bullet.

"He wasn't dying, honey," I said. "He was sunbathing."

One sunny morning I walked out onto the deck just in time to see a broad-winged hawk fly over my head and into the nearby spruce tree. Null, the smaller grackle, burst out of the tree like a racehorse from a starting gate and flew toward the field; close behind her was the broadwing, and running a distant third was me, screaming "No! No!" and shredding my own reputation as a staunch defender of the natural order. I stopped at the edge of the field, desperately searching the trees for signs of them, and found nothing.

"Where's Null?" asked Skye that night at dusk. "Look—Void is over there by himself."

"I don't know," I said. "But I'm sure she's all right."

I wasn't sure of any such thing, and I told John so. "What if the broadwing caught her?" I said. "What will happen to Void? He'll be all alone. There aren't any other grackles around here. What if the broadwing comes back? Do you think it's the same one that went after Zack?"

"Maybe if he does come back you should ask him," said John, adopting an exaggeratedly British accent. "Excuse me there, good sir, but might you be the same broadwing who, several years ago, attempted to eat our macaw? And have you, in fact, recently dined on our grackle?"

I laughed in spite of myself. "I can't stand this," I said.

The next morning Void flew in for breakfast, but there was no sign of Null. We spotted the robins, and, eventually, all the jays, but not the little grackle. With a sinking feeling in my stomach, I started to go inside.

"What's that sound?" said Mac. "It sounds like grackles."

I looked up. Soaring toward us, flanked by two wild adult grackles, was Null. I had never seen a grackle around our house in the summer, although I saw them regularly at the horse farm a quarter mile down the road. The small squadron flew over our heads and landed in the big spruce. When Null fluttered down to the deck railing I resisted the urge to grab her and give her a big kiss on the beak; instead I stared at the wild grackles, who were peering down at us.

"She must have gotten lost," crowed Skye, "and her new friends brought her home! How cool is that?"

"It's very cool," I said, a thousand questions—none of which would ever be answered—swirling around my head. Null strutted around the deck for a few moments, then joined Void at the feeding plate under the hemlock. The adults waited until she started to splash around in the water dish, then the two of them flew away in the direction in which they had come.

"Wait!" I called after them. "Don't go!"

"Maybe they're like those friends you have from New York City," said Mac. "They like being around kids for a while, but then they get fed up and have to go home."

The broadwing didn't return. The fledglings continued to grow and mature, to explore and discover. One day Void discovered anting, although I wasn't quite sure how he had discovered it. Anting is done by crows, jays, and grackles, as well as a number of songbird species. The birds either pick up single ants and place them in their feathers, or actually lie on anthills until the ants swarm all over them. The most common explanation is that the formic acid contained in the ants' defensive secretions kill parasites, like fleas and feather lice, which may be hiding in the birds' plumage. Some say that ants' secretions contain fungicidal properties, which would be helpful during humid summer months. Still others believe it's a vice, like drinking or smoking, that feels good but has no real purpose. At that point my only anting experience had been with crows, whom I firmly believed were in it for the cheap thrill and not the medical benefits; since I had no experience with jays and grackles, I wasn't yet sure about their angle.

In any case, one morning I went out to the deck and saw Void standing next to the food dish, picking up single ants and methodically placing them on various spots on his body. Standing next to him, watching carefully, was Null. Picking up a mealworm from the food dish she attempted to duplicate her compatriot's actions, but the worm would not cooperate. She placed it in the feathers on the nape of her neck, but the worm slid down her back; she picked it up again and placed it under her wing, where it dropped to the ground. I quickly called the kids, held my finger to my lips, and the three of us sat silently on the deck, watching as Null tried to figure it out. Finally Skye couldn't take it anymore.

"What are you doing, you crazy bird?" she hissed in frustration. "You don't use a *worm*, you use an *ant!*"

"That's why they call it 'anting' and not 'worming,'" added Mac. "Not that a bird would care what the word is in English."

Day by day the blue jays lost their fuzzy adolescence and became astoundingly beautiful young birds, bursting through the foliage with the bravado of musketeers, their crests raised at rakish angles. Everything seemed to be going according to plan, but our bright and noisy carnival had not gone unnoticed. Ten days after I released them I was sitting in the kitchen on the phone, looking out the window at the four who had gathered on a patch of ground. Suddenly a Cooper's hawk shot out of the woods, grabbed one of the little jays, and continued on out of sight.

I had never believed people who described an accident as time standing still, but I remember that moment as a photograph: the jay's dark eyes and the hawk's knife-edged intensity, a colorful young bird gripped by a sleek and lethal predator. I dropped the telephone and tore out of the house, but the other jays had scattered and there was no sign of the hawk.

When John came home from running errands I met him at the door, my eyes red. "A Cooper's hawk got one of the blue jays," I sobbed. "I saw it happen. I don't even know who it was—maybe it was Norbert. Or Hagrid. I don't know."

"Oh no," he said. "I'm so sorry. Where are the kids? Do they know?"

"They're not here. I can't tell them. I know I should be teaching them all about nature, even the hard parts, but they raised those blue jays and I can't tell them."

Later I sat on the deck stairs, watching as one of the robins pulled a worm from the bug pit. I had known from the beginning that my own silly human rules had no place in a wild bird's world. Still I had longed to be a part of it, and now that I was a part of it I was undone by one of its most basic laws.

Yet I couldn't curse the hawk. I would have been just as distraught if I'd released a young Cooper's and later watched him drop from a tree and die of starvation. "Whose side are you on?" my kids would have asked. Both, I would have answered. I'm on both sides, which means whatever happens, I can never really win.

"Where are the blue jays?" asked Mac, when I brought the kids back from their playdates.

"There's one," I said, pointing.

"That's Ron," said Mac. "What's the matter with him? He's acting kind of weird."

"Maybe something scared him," I said.

The hawk cut through the yard again late that afternoon, but the little group was not at their regular gathering spot. "Do you think he's after the fledglings?" asked Skye, and the apprehension in her face made me vow once again not to tell her what happened. The best I could do was to bring up the possibility.

"The hawk might catch one of them," I said. "It's a danger all wild birds have to face. But the hawk has to eat."

From then on the jays would appear one or two at a time, eat a few bites from the plate, and take off again, but they no longer responded to our loud announcements of fresh food. The grackles and the robins were more confident, but as the summer wound down they too spent more and more time away from the yard. One day three wild adult blue jays came to the feeder, and then days went by without any sightings at all. "Maybe Harry and the gang went off with the grown-ups," I said, not very convincingly.

I released the others, each time torn between the conflicting desires of protecting the fragile creatures I had come to love and experiencing the secondhand exhilaration of a just-released wild bird. As long as they were capable of living in the wild, there was no decision to be made; I could no more sentence a releasable bird to life in a cage than I could choose to dwell there myself. Yet that didn't stop me from agonizing over their potential encounters with predators, windows, cars, airplanes, pesticides, and plastic fencing, as well as excess heat, cold, rain, snow, fog, and wind. Not to mention possible bouts of loneliness. Or angst.

Romeo's tattered feathers molted and were replaced by beautiful new ones. The little catbird was fast and agile and could easily catch the live moths I trapped each night and released in the flight cage the next morning. I drove him to a perfect spot, a marsh by the Hudson River filled with pokeberries and other catbirds, and opened the carrier. Romeo bolted out and I felt a rush of air, the upward sweep of a long-ago rope swing, and just for a moment I thought I could follow him into the towering oak tree. Then the leaves closed around him and he was gone.

I released Orangina, the finches, and the wrens in the backyard, hoping that the single cardinal would join the resident pair who appeared at the feeder several times a day. Before the week was out I saw Orangina land on a branch right next to the wild male—her beak open, wings vibrating so violently I thought she'd fall from the tree. For the rest of the day I'd see the pair of wild cardinals racing around the house, frantically pursued by the furiously begging Orangina. By the following day they had bowed to the stronger will and were taking turns feeding her, and the faithful pair became a trio.

The waxwings had worked it out. Yin had pursued Yang with a determination so dogged that within a few days Yang was not only allowing her to perch next to him, but on occasion would also ceremoniously feed her a mealworm or

a blueberry. Waxwings live in large flocks, and in order to give a captive-raised waxwing a halfway decent chance of survival you must release it directly into a flock of wild ones. Thanks to my friend Lew Kingsley, who always knows where the birds are, we accomplished this needle-in-a-haystack feat, and Ying and Yang were absorbed into a welcoming flock.

I found out where the shy, secretive wood thrush had come from, drove him to a forested area two towns away, and let him go. I released the downy woodpecker in the yard, as I kept two suet feeders full, the woods were filled with dead hemlocks, and there were other downies in the area. Last to go was the herring gull.

"I have this vision of releasing the gull on a beach in Nantucket," I wrote to Ed, "not in a parking lot next to some disgusting burger joint."

"If you released him on a beach in Nantucket," Ed replied, "odds are he would immediately start looking for a parking lot next to some disgusting burger joint."

The kids and I drove the gull back to Burger King on a warm, overcast Saturday morning. When I tossed him into the air he flew strongly upward, made two large circles, and coasted down, landing easily next to the small group of standing gulls. They all gazed at each other and rustled their wings; our gull approached another and gave him a sharp poke with his beak.

"See?" said Mac. "It wasn't just you."

A car approached, and the group took off into the air. I still wanted them to be skimming over the sea, turning their backs to a gathering nor'easter, and searching the beach for crabs and dead fish. But these aren't the ones I should be worrying about, I thought. The aggressive opportunists are doing just fine. The ones in trouble are the ones no one will ever see in a Burger King parking lot: the black-capped vireo, the piping plover, the golden-cheeked warbler, the spectacled eider, the whooping crane, the California condor, the bobolink, the meadowlark.

There were so few of them left, and I had just saved a common herring gull. I looked up at the circling gulls, who flew just as gracefully over a parking lot as a condor over a valley, a warbler through a forest, or an eider above a lake. Common or not, I could never begrudge a bird its wings. Should someone bring me an injured black-capped vireo, I decided, I would do my best to save her, too.

Even though I wasn't taking any injured birds.

        ⊚   ⊚   ⊚   ⊚   ⊚

The house finch had been my first songbird, and it was he who was left behind at the summer's end. Although there was no infection, his eye had begun to atrophy; his wing had healed, but not well enough for life in the wild. He had a firm place in my heart. In his own energetic, matter-of-fact way he had dealt with injury and captivity, and had been the supervising adult for more than a dozen fledglings. His close companion was a young female orchard oriole, a lovely bird whose broken wing had not healed well enough for her to ever fly again. Her right wing drooped noticeably, but she was alert and active, and she hopped around the entire flight cage. Unfortunately, I did not have the permits or the facilities to keep unreleasable birds; my rehabilitator's license stated that if I couldn't find a good sanctuary with proper permits willing to accept them, I would have to have them euthanized.

"I'll find them a good home," I said to John, adding silently, even if I have to drive them to California to do it.

I was the den mother whose den had emptied but who still could not relinquish her hold on her scouts. I thought about the released birds constantly. I worried about them. I was able to keep tabs on some of them, for the birds in our yard were no longer anonymous. When we looked out the kitchen window, it was like looking onto the main street of a small town. Sometimes we'd see a familiar little group of finches, sometimes a robin poking through the bug pit. A trio of cardinals. Grackles perched in the black oak. A downy woodpecker swinging from a suet holder.

I took stock. Despite the work, the stress, and the constant juggling, I felt as if I was where I was meant to be. After my first summer of home rehab, I couldn't imagine a life outside the bird carousel.

We continued to look for Harry Potter and his siblings, but caught only glimpses of them from time to time. Then one morning John went running and took the trail through a wooded area by a nearby pond, christened "Blue Jay Town" by the kids for its large and noisy population. Hearing a familiar sound he stopped, then looked up to see a small group peering down at him from the top of a tree. He shouted his usual "Hey, you blue jays!" and two of them hopped down for a closer look—something a born-wild jay would rarely do—then they all flew off together. The next day he returned and saw a young one settled firmly on a branch, being fed by another.

"It was Norbert!" said Skye delightedly. "I knew it! They've all moved up to Blue Jay Town, and they're going to live happily ever after."

In a world filled with all kinds of possibilities, I had no reason to doubt her.

PART TWO

Chapter 21

BRANCHING OUT

"The femur is fractured," said Denise. "I had him X-rayed, and it needs to be pinned. But the only vet who can do the surgery is on vacation, and I've called everyone else I know. Do you have a vet who could do it? One of my volunteers could drive the bird right down to you."

"But Denise!" I said. "I've never taken care of a heron!"

"Quiet, no stress, feed him live fish," said Denise. "Wear goggles so he can't poke your eyes out."

The kids had gone back to school. Free all summer, they were now enclosed; with no more baby birds to feed, I had been set free.

But not for long.

Denise Edelson is a long-time rehabilitator in Woodstock, New York. She accepts all kinds of injured wildlife, from opossums to birds to turtles. Whenever she can find a weekend babysitter for her animals she and her husband take off into the Adirondacks to study loons. She has a softly memorable voice and nothing fazes her.

"Let me see what I can do." I sighed.

Alan was on vacation. Wendy was not due to work at the hospital for three days. A reknowned orthopedist an hour away was booked solid for surgery. I called Maggie.

"Can you rehab a heron?" I asked her.

Maggie burst out laughing. "A heron! Right! Good luck finding someone to take a heron!" she said. "But as far as vets go, try Croton Animal Hospital. Dr. Hoskins and Dr. Popolow are wonderful, and they've always helped me out."

Dr. Popolow agreed to do the surgery, but the hospital's surgery day was not until Thursday—two days away. I called Denise back, opened my purple three-ring binder, and looked up "Great Blue Herons."

I had no business taking this bird, as I didn't have the experience or the right facilities. I could drive him to a rehab center several hours south, but I had to be here when the school bus arrived. It seemed as if I were the heron's only chance. I read through the heron information twice, made a checklist, and went to work.

By the time the heron arrived I had dropped his X-ray off at Dr. Popolow's hospital, bought a few dozen live fish, and outfitted the shower in my own bathroom with five-foot phragmites freshly cut from a marsh near my house. My bathroom was quieter and more isolated from the rest of the house than the bathroom in which I had put the gull and the nestlings, and I wanted the stress-prone heron to be as free from disturbance as possible. I fixed the tall reeds to the shower walls with clear packing tape, put rubber mats on the shower floor so he wouldn't slip, and placed a green rubber tub containing a dozen live fish on top of the mat. I plugged our sound machine into the wall and turned it to "Babbling Brook," then went to dig my ski goggles out of the closet in the garage.

Soon after, I was walking through the house carrying a very large and aggravated heron. The adult great blue is four feet tall, has a six- to seven-foot wingspan, and sports a sharp and powerful beak. My heron information included a diagram showing how to hold the bird so as not to end up with this legendary weapon imbedded in one's eye: one arm encircling the body, the opposite hand firmly grasping the beak. Using this technique I gently put the heron into the shower, noting that he was strong and in good weight. I also noted that he was staggeringly beautiful; a Giacometti with a sword, hard-eyed and elegantly plumed. Hoping the phragmites made the area look less frightening, I closed the shower door and hurried from the room.

One spring day when I lived in Maine I followed an abandoned logging trail deep into the woods. I heard a series of harsh, guttural croaks and, searching for the source, finally raised my eyes to the tops of the pine trees. Over a hundred feet in the air and silhouetted against the sky were dozens of great blue herons, all guarding huge nests filled with downy chicks, in a scene that struck me as strangely primordial even before I learned that herons (as well as egrets and bitterns, their fellow Ardeidae) have remained essentially unchanged for the last 1.8 million years. They were like a lost order of Druids, ancient and wise, an integral part of the earth and sky and far beyond the understanding of modern humans.

And now I had one in my bathroom.

A relative silence descended on the house. "No running through the living room. No yelling. No going into our bedroom. No loud TV," I said during dinner.

"Can we breathe?" asked John.

"Only through your noses," I said.

The following morning his fish were uneaten, a few still swimming under the water but most floating on the surface. Donning my goggles I held the heron, pried open his beak, and slipped several fish down his throat, making sure each one made its way down his long neck. I left quickly, giving him less of an excuse to immediately regurgitate them.

I went outside to feed the house finch and the orchard oriole. Soon the weather would start to turn cold, and I needed to find the two of them a good home. I put a message on the Wildlife Rehab electronic mailing list, and within a few hours had a reply from Leslie Hayhurst, a mailing-list regular and the founder of Genesis Wildlife Sanctuary.

"Sure, I'll take them," she wrote. "We just built a beautiful new songbird aviary—it's climate-controlled and filled with live trees and has lots of ramps for the ones who can't fly. They'll have a good life. All you have to do is get them here."

"Here" was the mountains of North Carolina. Well, I thought, I had vowed to drive the house finch to California, so it could be worse. I put another message on the mailing list: anyone happen to be leaving New England and heading for North Carolina? No replies.

That night I was reading in bed when John came in. "Have you noticed . . ." he began, then stopped when I grimaced wildly and stabbed my finger toward the bathroom.

"Have you noticed," he whispered, "that it smells like Sea World in here?"

"I know," I whispered back. "But I don't have anywhere else to put him. He needs his own building."

"How long will it take him to recover after the surgery?"

"A few weeks."

"Hmmm," he said. "This could get interesting."

"You're a good egg," I said.

"I know I am," he replied. "Just don't put that dinosaur in my office."

The next morning John and the kids stood in the living room as I carried the heron past them and into the garage, where I would put him in a carrier and drive him to the animal hospital. It was their first sighting, as I had been too worried about his stress level to allow anyone else into the bathroom.

"Wow!" I heard softly. "Look how cool he is!"

At the hospital, Dr. Carol Popolow greeted me with a direct blue gaze and a wide smile, her obvious compassion and enthusiasm for her work instantly putting me at ease. As I would eventually learn, Dr. Popolow will efficiently assess a bird's injury and its chance of recovery, weigh the amount of stress involved in its treatment against the likelihood of its release, and lay it all out in a clear and understandable way. I left the hospital filled with happiness. I had survived my first summer of solo bird rehab, I had found another veterinarian willing to help me, and I was infatuated with great blue herons. On the way home I was thinking about what our bedroom would smell like after several weeks of heron rehab when I noticed a garden supply company, in front of which sat several barn-shaped garden sheds.

"Sure, I guess," said the owner. "You could drywall it and paint it, then all you'd have to do is run a line up through the floorboards for electricity. Install some lights, and get one of those heaters where you can set the temperature. Where do you live? Yeah, we could deliver it."

The more I thought about it, the more sense it made. How could I keep the kids quiet for weeks at a time? What if another great blue heron was injured, and there was no one to take it but me? What if someone brought me a killdeer—didn't they also need a perfectly quiet environment away from humans? What did killdeer eat, anyway?

The summer had been expensive, and live fish were not cheap. An 8- by 12-foot garden shed cost a lot of money. I currently had no income and my husband was a freelance writer. How would I pay for everything? This one didn't take long to figure out. I had written successful fund-raising newsletters, I could write one for myself. I could set up a nonprofit corporation. I could write the newsletter during the winter, when things were slower.

I didn't feel myself falling.

It was hard enough for me to believe that great blue herons existed at all, let alone that I had actually kept one in my shower, carried him to the vet, and would be helping him to heal. I was baffled that it had taken me this long to come into physical contact with a creature so awe-inspiring. In a just world there would be a wildlife rehabilitator on every corner, but in this world I had been the heron's only chance. What about all the other birds, for whom I might be *their* only chance? The situation was so unfair. But it could be remedied: all I needed to do was break a few more of my own rules.

I sat down and made lists. (1) People to call for nonprofit advice. (2) People to call for newsletter-publishing advice. (3) People to call for drywalling advice. I returned to my computer and checked my e-mail, where I had no offers to drive a group of unreleasable songbirds to North Carolina. I opened my purple three-ring binder and reread my heron chapter, then opened my yellow three-ring binder and reread my chapter on "Post-Surgical Care." Just before the kids were due home from school, the phone rang.

"Suzie?" said the voice. "It's Carol Popolow. Hi. Listen, I'm afraid I have some bad news."

But that's impossible, I thought. How can you have bad news?

"Too much time had elapsed between the injury and surgery, and his ligaments had tightened," she said. "They pulled the bone so far apart that I couldn't get the two pieces close enough to pin. I pulled with all my strength and I couldn't get them together. I'm sorry—he was such a beautiful bird. But they can't live on one leg. I euthanized him while he was still under anesthesia."

I had been unprepared for anything but resounding success. "Thank you," I stammered. "Thank you for everything you did for him. I . . . I'm really sorry."

"Me, too," she said. "I wish we could have gotten him into surgery faster. I'll try again, if the situation comes up."

I hung up the telephone and the heron flew by the kitchen window, graceful and sinuous, the sound of his wings as strong and steady as the echo of a bass drum, so real that it was only after I watched him go that I realized he had never been there at all. The ring of the telephone made me jump.

"Suzie? It's Jen Bowman. Remember? I brought you the grackle and the waxwing. I heard you have birds who need a ride to North Carolina. My husband Wendell and I love road trips—we'll take them and make a long weekend out of it. I have three unreleasables here at the hospital, and I called Leslie and she said she'd take all five of them. We'll spend Saturday night at her house. She's such a nice woman—this is really going to be fun."

When the kids hopped down from the school bus I smiled at them, but Mac wasn't buying it. "What's the matter?" he said. "What's wrong?"

"Is it the heron?" asked Skye, looking apprehensive. "Is he dead?"

"Yes," I said. "He is. He didn't make it through surgery."

"Ohhh," said Mac, watching me closely. "I'm sorry."

"Oh, no!" cried Skye, then saw me blinking rapidly and abruptly switched gears. Linking my arm in hers, she leaned her head against me. "It's going to be okay," she said.

"Thanks. Thank you both," I said. "You know, I found a ride for the finch and the oriole."

"Who's taking them? When are they going?" they asked, peppering me with questions. We walked up the hill to the house, three watchful, careful guardians of one another, each taking a turn being the shoulder to lean on.

Chapter 22
DIVING IN

For a week I had no wild birds, and my list of people to call for advice sat idle while I did all the chores that had been pushed aside by the hectic summer. I had been so dazzled by the heron that his death had thrown me, yet I found myself listening for the ring of the telephone. Although I had made no announcement, I had crossed the divide. There were so many injured birds in the world—didn't any of them need my help?

"I hope you can help me," said the voice on the telephone. "Maggie gave me your number. I have an injured peregrine falcon."

I took a quick breath. Peregrine falcons are the jet fighters of the bird world, dramatically colored little self-guided missiles that disappear into the sky and then knife downward at their prey at a heart-stopping 200 miles per hour. Their name comes from the Latin word for "wanderer," as their migrations take them to both hemispheres. Nearly driven to extinction by DDT and other pesticides, they have made a remarkable comeback but are still on the Endangered Species list in New York as well as in twelve out of eighteen other eastern states. Those who work with raptors have their personal favorites, but all are awed by the peregrine, just as all sports buffs are awed by the athlete who possesses the perfect combination of muscle, skill, and heart.

A second later I let my breath out and smiled. Rehabilitators love to talk about their cases of mistaken identities. An "opossum" turns out to be a rat, a "baby eagle" a pigeon, a "rabid hawk" an angry duck. My favorite was the goodhearted Maine tourist who chased a "deformed Great Dane" through the woods, determined to get it some medical help, until she was interrupted by its irritated mother (a moose). "Peregrine falcons" nearly always turn out to be a Cooper's or a sharp-shinned hawk, if not a crow or a grouse.

But whatever it was, it needed help.

"The problem is, I can't get him to you," continued the voice. "Is there any way you could come and pick him up? I'm at Entergy."

We are not too far from Entergy, better known as Indian Point. After 9/11 there were massive protests and demands to close the nuclear power facility, which is located on the Hudson River only fifty miles north of New York City, because many nearby residents feared it could be a terrorist target. The company held firm, adding more security, widening its PR campaigns, and eventually riding out the protests.

The bus arrived, and Mac volunteered to come with me while Skye stayed with John. I explained the fact that "peregrines" were never peregrines, and we took turns guessing what the mystery creature would be, each guess becoming more and more outlandish.

"It's a naked mole rat!" I said.

"It's an Australian frilled lizard!" countered Mac.

Finally we drove past the ENTERGY sign and caught sight of five burly men, all holding very large machine guns. It looked as though the newly ramped-up security system was working, at least around the entrance booth. Each man turned, scowling, and stared at us. Mac gasped.

"Mom," he whispered. "This is *so cool*."

It didn't seem like the right time to start a discussion about either arms control or the downside to nuclear power, so I kept silent as I pulled up to the booth. One of the men leaned, unsmiling, toward the car. "Help you?" he snapped.

"I'm, uh . . . here to pick up the . . uh . . . bird," I stammered.

Instantly the man broke into a wide grin. "Oh! The bird!" he said. "Come on in!" He gestured to the others. "She's picking up the bird!"

"They're here for the bird!" chortled one of the men, slinging his gun into a more casual position over his shoulder.

"That's a really nice bird," replied a cohort, adjusting his bulletproof vest and nodding vigorously.

"Mom!" Mac hissed. "You think you could get them to shoot something for me? Like how about that car over there?"

"Here's the patient," announced yet another man, who had materialized holding a cardboard box. In short order the small army opened the back of my Jeep, placed the box carefully onto the floor, and quietly closed the door. "Take good care of him," rumbled the one who'd brought the box, while indicating that I should turn the car around and leave without delay.

"Awwww, we can't leave yet!" Mac protested on the way out. "Say, how old would I have to be to get one of those guns?"

"Eight hundred and six," I replied, "and don't think I'm kidding, either."

It was a Friday afternoon, Wendy was at the hospital, and her office hours were nearly over. In short order Wendy, Mac, and I were standing around an examination table, upon which rested the mysterious box. "It's a 'peregrine falcon,'" I said, hooking my fingers dramatically around the name.

I opened the box. We all stared, open-mouthed.

"Whoa!" said Mac. "It's a peregrine falcon!"

A wild peregrine is the embodiment of speed and unfettered flight; to see one injured and grounded is to be affected not only by the individual bird but also by the stark symbolism of its condition. Our falcon huddled in the box, too weak to move; infinitely precious, and almost dead.

We guessed that the young bird, less than a year old, had either flown into an Indian Point window or been blown into one of the nearby Hudson River bridges by a sudden downdraft. His beak was split from the nares (nostrils) down to the tip, which had separated into two fanglike points. The surrounding tissue was black and caked with flecks of dried blood. More alarming, though, was his weight. His normally strong, firm chest muscles had dwindled as starvation consumed them; now, in the final stages of emaciation, they had all but disappeared, leaving nothing but skin and bone.

Dehydration could be remedied with subcutaneous fluids, but reintroducing food into his depleted system would be trickier. At this point solid food would kill him, as his body would use the last of its fading reserves in a futile attempt at digestion. A liquid, easily absorbed food mixture would have to be syringed directly into his crop—birds' temporary food storage area, located at the base of their esophagus—by inserting a tube down his throat. Only when his system was working again would he be capable of eating solid food.

"You still not taking injured birds?" asked Wendy with a grin.

"I'm being sucked into the abyss," I said.

"Join the crowd," she said. "You'll have to feed him every two hours for the next couple of days."

"That's all right," I said. "How late into the night?"

"How long can you stay awake?" she replied.

It wasn't difficult to tube the peregrine, as he was too weak to resist. I wrapped him like a papoose, straightened his neck, opened his beak, inserted the tube, and before he knew it, it was over. That night I gave him his midnight meal, yawning, then set him up in a medium-size crate in John's office, thinking it would be quieter. I turned the heat up to 80 degrees, placed a heating pad in the back corner of the crate, draped it with blankets, and turned off the lights.

The next morning at 6:00 I hurried into the office, looked into the crate, and recoiled. The little falcon was lying on his chest, breathing heavily. When I reached in to pick him up his eyes closed, his breathing stopped, and he was still.

I stared down in disbelief. My first Endangered Species was dead. How could this have happened? Why did I ever go bed? And what if the Indian Point Army found out I killed their bird?

I grabbed a blanket from the top of the crate, wrapped it around the peregrine, and cradled him. "You can't die," I said pathetically. "Wake up. Please. Wake up."

There was no response. I tucked the little bundle against me, searching in vain for any sign of life. Finally, with an ache in my throat, I started to put him down. As soon as I did, his head moved. His eyelids fluttered, and slowly he opened his eyes.

Later that morning, after he had been moved into the house and was thoroughly warm, stable, and fed, I sat near the crate, feeling like I'd survived a mortar attack.

"How's the patient?" said John, coming into the room.

"I'm trying to figure out exactly what happened," I said. "Physiologically."

"Maybe he saw The Light and flew the other way," said John, who feels compelled to provide some sort of response to all unanswered questions. "Maybe he suddenly saw all his dead relatives staring at him and decided to return to your loving arms."

"Will you please get me another towel?" I said, with irritation.

For the next two days I stuck to the schedule like an overzealous new mother, warming the mixture to the perfect temperature and delivering the goods every two hours from 7 A.M. to 10 P.M. Unwilling to house him more than a few steps away from me, I put his crate in our bathroom. Occasionally Mac and Skye would look in, exclaim over the general awesomeness of peregrines, and then disappear, graciously accepting the new household rule that kids should not pal around with recuperating raptors.

One afternoon I was standing in front of the kitchen sink, concocting more tubing mixture and studiously ignoring the telephone, when I heard a commotion coming through the machine.

"*Ciao, bella!*" the voice shouted. "Put the bird down and pick up the phone! I know you're in there!"

"*Ciao, bella!*" Mario roared back in Italian. "*Bel uccello!*" (Beautiful bird!)

It was my old college buddy Pancho Castanheira, the crazy multilingual

Brazilian Buddhist who had lived a peregrine-like existence before finally settling in San Francisco. When we first met we spent the evening comparing the schools and clubs we'd been thrown out of and drinking the aptly named Scaglione, a red wine guaranteed to take the paint off a car. "You got kicked out of *military school*?" I remember saying. "Now, that's impressive."

Pancho loves peregrine falcons and was delighted to hear that I was hosting one. "Did you give him a name?" he asked. "Pancho," I said, without hesitation. "Junior."

Pancho Senior supplied me with two phrases in Spanish, his first language, and from then on whenever I approached the peregrine I'd repeat them softly: "*¡Hola, Panchito!—¿Cómo está mi pájaro guapo? ¿Te sientes mejor?*" ("Hello, little Pancho—How is my handsome bird? Are you feeling better?") Eventually, whenever I would say those words, he'd cock his head to the side, which I translated from the peregrinese as "Yo—Anything good on the menu?"

By the third day the menu included some solid food. One of the complications of rehabbing birds of prey is that they need to eat entire animals—organs, skin, bones, and all—to stay healthy. The charming story of someone's finding a baby owl in the woods and raising it on hamburger never seems to include the ending, when the owl develops metabolic bone disease and eventually dies of calcium and other vitamin deficiencies. Anyone feeding a raptor properly needs to have a freezer filled with things that would make the average person scream and run.

At this point Pancho's delicate system was not ready for a whole animal, so I dug through the freezer until I found a bag labeled "small mice," given to me over the summer by a rehabber dropping off a songbird. I put one mouse into a baggie and then into a container of hot water, and a half hour later filleted it into a few prime bites. After his tubing mixture I opened Pancho's beak and put in two bites of mouse, which he was perfectly happy to swallow.

Most birds of prey are solitary creatures, and they value their freedom. They are not touchy-feely, like parrots; they will tolerate a human being as long as the human brings them food and doesn't annoy them, but in most cases if the situation presents itself they will take off without a backward glance and never return. Young raptors, however, especially young ones recovering from a catastrophic injury, can become quite relaxed and comfortable around their initial caregiver. In the short term, this is a good thing, as any kind of stress will hamper healing. In the long term, however, it is not a good thing; you cannot release a raptor that will land on an unsuspecting person's head the minute he feels a hunger pang. Once the bird is active and healthy, the best thing to do is either put him into a flight cage, where he will soon resume his independent ways, or transfer him to another rehabber. Or both.

At this point Pancho was far from healthy, although he was stable and quickly gaining strength. I took him back to Wendy for a checkup, this time transporting him in a carrier with a perch.

"What a difference!" said Wendy. "He looks so much better."

Beaks and fingernails are made from a substance called keratin, and both grow the same way. Only time would tell if the growth plate behind the falcon's beak had been damaged beyond its ability to heal itself. Meanwhile, Wendy tied the split ends together with catgut and then epoxied the middle, hoping to stabilize it enough to allow him eventually to eat by himself. It was a definite improvement, of course, but with his huge dark eyes, beautiful coloring, and battered beak, he reminded me of a spectacularly handsome hockey player who smiles to reveal a mouthful of broken teeth.

The regenerative power of wild birds can be truly remarkable. Six days after nearly starving to death, Pancho was hopping from log to log on the floor of the bathroom, which I'd set up to give him more space, while clouds of steam loosened the dried blood from his nasal passages. No longer on a liquid diet, he'd snatch gory little mouse bits from my oversized tweezers as his weight continued to climb. My laundry area became a rendering plant; the kids would walk up the hill from the school bus and find me hunched over the washing machine like a demented sushi chef, animatedly skinning small piles of rodents.

"She's doing that mouse thing again," Mac would say with a doleful sigh.

"Eeewwww, Mommy, you are *so gross*," Skye added disgustedly.

I called Paul Kupchok, the naturalist, educator, and rehabilitator who, back before his semi-retirement, ran the wildlife program at Green Chimneys School. One of the tenets of this widely heralded nonprofit organization in Brewster, New York, is that children with emotional, behavioral, or learning difficulties can benefit from animal-assisted therapy. The school has farm animals from all over the world and a renowned raptor program. They also have large raptor enclosures—which I don't have.

"He's fine for now," I said to Paul, "but pretty soon he's going to be bouncing off the walls. Would you be able to take him then?"

"Sure," said Paul. "I love peregrines. You called Albany?"

When a New York State rehabilitator receives a member of a state-listed Endangered Species, the rehabber must contact the Endangered Species Unit of the New York State Department of Environmental Conservation (NYSDEC) and let them know. If the bird is listed as Federally Endangered, the rehabber must also contact the U.S. Fish & Wildlife (USF&W) Service. The NYSDEC issues state licenses and USF&W issues federal permits; ignoring either of these agencies, or even worse, making them mad, can result in dire consequences for the rehabber.

"Ahhhhhh, jeeeeeez," I croaked. "I forgot. I have a week, don't I?"

"You're in trouble!" said Paul in a cheerful voice.

Peregrines had been taken off the Federal Endangered Species List, but were still listed as endangered in New York, so in short order I was on the phone with Barbara Loucks of the NYSDEC Endangered Species Unit.

"I thought I had a week to call," I finished lamely.

"A week!" she said. "Try forty-eight hours!"

I stopped: the red flag was waving in front of the bull. I was being dissed by an Authority Figure, my nemesis since childhood. Should I have spent the past week reading the fine print on my #@*!%* license, I wanted to shout, or making sure the damned bird didn't die?

I took a deep breath. "I'm sorry," I said. And oddly enough, I actually meant it. The peregrine currently residing in my bathroom was a genie from a bottle, a phoenix who had crashed to earth and risen again, a wild creature of such spellbinding beauty and ability that merely being in his presence took my breath away. I knew he couldn't stay here for long, but I suddenly realized that I would say or do anything to guarantee those few extra days in his company.

"Suck-up!" screamed the little voice in my head.

Chapter 23

LOOKING UP

I had to have better facilities. What if another peregrine was injured, and there was no one to take it but me, and the bathroom was already occupied?

"John," I said. "I need to buy one of those garden sheds so I can turn it into a clinic and get the birds out of the house."

"Really," he said, looking pained.

"I need to pay for it up front, then I'm setting up a nonprofit and asking for donations."

"But what happened to the 'no injured birds' rule?"

"I don't know," I said. "It came crashing down, like the Berlin Wall. But, really, it will be fine—it will be a small clinic, and I'm only one person so I can do only so much. I can't take in that many birds. Think about it—isn't that peregrine the most amazing creature you've ever seen? And what would have happened to him if I hadn't been here?"

We tramped around outside, trying to figure out where the best spot for the shed would be, settling on the area between John's office and the flight cage. If the shed were placed back-to-back with the office, we could run a thick extension cord from the office's outside outlet and up into the shed. The shed would be in a quiet area away from the house and only a few steps from the flight cage. There was only one problem. "How are we going to get it back here?" asked John.

"I'll figure it all out," I said brightly. I returned to the house, picked up the phone, and started dialing.

First, I called Lew Kingsley, arborist, naturalist, veteran birder, and local curmudgeon, who always knows where the birds are and had located the wax-wing flock into which we released Yin and Yang.

"What do you want now?" he said. "More cedar waxwings?"

I described the situation. "Oh, fine," he said, "but how do you think you're going to get it back there?"

"Lew!" I exclaimed. "That's what I'm calling you for!"

"Oh! Right. Well, you get some big rollers and a bunch of pulleys. You put the rollers under the shed, and you run a rope around it and then around that hemlock by the parrot cage, and you just pull it forward. Then you run the rope around that big oak by John's office, and you pull *that* forward. Just keep inching it forward. It'll work."

"Can you help me, if I get the shed here in a few days?"

"Sure. No problem."

Next, I called John Cronin, once the Hudson Riverkeeper, a long-time environmental crusader wise to the ways of nonprofits. "It's such a tiny operation," I told him. "It's just me. I don't want to expand. Is it worth it to try to set up a 501c3?"

"Yes," said John, who neither minces words nor sits on the fence. "Absolutely. It's not as hard as you think, but you need a lawyer to help you set it up. Do you know Bob Bickford?"

I called Bob Bickford, a retired lawyer known for his generosity and community-minded spirit. "That sounds like a wonderful idea," said Bob. "Sure, I'll help you. I'll e-mail you some material. You need to pick a board of directors, usually five to seven people."

"Would you be on my board of directors?" I asked.

"I could do that," he said agreeably.

I called Randi Schlesinger, a graphic designer who lives at the horse farm down the road. Randi not only has a gift for finding the most arresting image-typeface combinations, but she is a dedicated greenie and animal lover and has the heart of a philanthropist.

"Randi!" I said. "Can I hire you? I'm setting up a nonprofit for my bird rehab operation and I need to create a newsletter. Something simple. I can supply the copy, but computers scare the life out of me."

"Wow!" said Randi. "That is so cool! I'll bet we can come up with something really eye-catching. I'll tell you what—you just pay for the printing and the rest I'll donate. Because it's a *really* good cause."

I called Mark Kristiansen, who can build or fix anything. "Sure, I could drywall a shed," he said. "Let's see. I could start a week from Wednesday. Good for you?"

I called the garden supply company and arranged to have the shed I wanted delivered. I checked off the list I'd made, started another, then served Pancho a quail (a peregrine delicacy), a frozen box of which I had ordered through the Internet.

After my conversation with the Endangered Species Unit I became a model rehabber, faithfully calling Barbara Loucks with weekly progress reports, feeling less like a suck-up after being roundly chewed out by a fellow rehabber.

"You didn't call DEC for a week?" she exclaimed. "What's the matter with you? It's an Endangered Species! They want the best possible care for the bird—if they don't know you, how can they be sure you're not going to put a feeding tube down his windpipe?"

When the epoxy on Pancho's beak flaked off and it split once again, I took digital photos of his operation and sent them to Barbara via e-mail. Along with colleague Dr. Mary Wallace and veterinary technician Shirley Fogelquist, Wendy put the peregrine under anesthesia and then put a hole through the top of the beak, where the tissue was dead, and inserted a wire. She twisted the wire enough to draw the two sides of the beak together, snipped off the ends of the wire, and covered the whole thing with epoxy. This time it held, and a week later Wendy added a bit more epoxy and filed it into a point, allowing Pancho to have a temporarily functional beak.

Meanwhile the patient continued to improve. No longer content to hop from log to log on the bathroom floor, he'd jump up onto the towel-covered toilet and eye the window ledge with more than casual interest. Convinced that he was destined to slip off the ledge, fall behind the toilet, and break his neck, I banished him from the bathroom floor and instead put him in an extra-large dog crate with several perches. This seemed to ratchet his energy level up several notches. Normally when I picked him up and placed him on the scale for his daily weight check he stood casually, wearing a slightly bemused expression. Now, as soon as I lifted him out of the crate he screamed at me in his goshawk-crossed-with-seagull voice and bicycled his legs so frantically I had to put him down on the floor. He'd glare at me, rattle his feathers defiantly, then trot over to the scale and hop up by himself.

The day he started eating from a dish instead of my tweezers I committed an act of hubris that even now makes me shudder. As I ambled through the house I actually thought to myself: *you're a pretty decent rehabber.* This is akin to spitting in the face of the rehabber gods, and it didn't take long for them to retaliate.

I had put the crate on my bed so that Pancho could get some afternoon sun. I walked into the bedroom and the crate door was ajar. I must not have locked the gate securely. Pancho was gone.

I turned cold. I imagined the peregrine knifing through the house, grabbing a parrot in each foot; I envisioned Mario begging for his life by singing "Rainy Night in Georgia" with extra pathos. I shot into the living room, completely distraught, and shouted to the kids.

"Oh, my God, Mac and Skye! Pancho's gone!"

"No, he's not," said Skye, without even looking up from her drawing. "He's sitting up on the bedroom door. I thought you were letting him have some exercise."

As I raced from the room it occurred to me that my seven-year-old daughter hadn't thought twice about finding a fairly large carnivorous bird loose in the house. My Parent-o-Meter took another nosedive.

I found Pancho perched contentedly on the top of the door. *"¿Cómo está mi pájaro guapo?"* I asked. He looked down at me and cocked his head: anything good on the menu?

The day the shed was delivered John and I watched as a large flatbed truck rumbled slowly up our driveway, did a three-point turn, and backed up toward the house. Two men eased the shed off the truck and onto the top of a small embankment, the very beginning of our lawn. Still to go was fifty feet across the lawn, over a small retaining wall onto a lower level, then 120 feet along a long, sloping, rocky path to the back of John's office.

I ran into the house to call Lew. "You're not going to believe this," he said, "but a big job I've been waiting two months for just came through. If you can wait about five days from now, I'll be there to help you."

I relayed the information to John. "I guess we can just leave it here until then," I said.

"It's supposed to start raining on Sunday," he said. "For a couple of days. It'll be a mudhole. We can't wait." He paused. "I'll do it."

The shed was eight feet wide by twelve feet long and nine feet high.

"What do you mean, you'll do it?" I demanded. *"It weighs two thousand pounds."*

It took him almost three days. John bought six round plastic pipes eight feet long to serve as rollers and a long, heavy pole to act as a lever. Our friend Robert helped him to get the shed over the retaining wall, but aside from that, John worked alone from 7 in the morning until 6 at night, sliding the shed forward inch by inch, refusing all offers of help. Occasionally he would appear in the doorway, covered with sweat and dirt, and ask me to lean on the shed while he slid another roller beneath it. I would finish my small task, then insist that he let me provide him with further assistance.

"Nope," he'd say. "Go away."

Halfway into the third day I followed him down the rocky path and there was the shed, nestled perfectly against the back of the office.

"There you go," he said.

As much as I wanted Pancho to be the shed's first guest, it was not to be. He had grown even stronger and more energetic, and the turning point came later that afternoon when I opened the crate door and he hopped out onto my arm. He looked me in the eye, then hopped up onto my head. "Hey, get off me!" I ordered, grabbing him and putting him back in the crate. "No bird hats."

He was ready for an outdoor flight, and he had become too attached to me. I called Paul. "It's time for him to go," I said, fighting the quaver in my voice. He had been with us for three weeks.

The next morning I told the kids I was taking him to Green Chimneys. "You mean no more mouse guts?" said Mac.

Paul had a small enclosure ready, built next to the side of a building and protected from the wind. It was beautifully constructed, with evenly spaced wooden slats and perches of varying height and dimension. When I took Pancho out of his carrier he surveyed his new surroundings, then stared intently into my face.

"Look at the way he looks at you," said Paul.

Paul kept me informed. He taught Pancho to wear jesses, the soft leather leg straps worn by falconry birds, and to perch quietly on his glove. He taught him to chase a lure swung on a long line, hoping that these falconry techniques would ready the bird for his eventual release by increasing his speed and agility. But although Pancho could swoop through the air in pursuit of the lure, he couldn't catch it; each time he would overshoot it, landing a foot or so beyond his target. Paul began to suspect that the blow that had split his beak had also damaged his eyesight.

Paul took him to an avian eye specialist. The exams proved inconclusive. After several months there was no sign of beak regrowth, so Paul called two of the country's top raptor veterinarians. They consulted with a friend of Paul's, a surgeon who worked at a local hospital, who then invited a dental surgeon to brainstorm whether it would be possible to create a prosthetic beak. The verdict was that no artificial substance could withstand the constant wear and tear normally endured by a raptor's beak.

I called Barbara with the news. Paul agreed to provide him with a permanent home.

⊚   ⊙   ⊚   ⊙   ⊚

A little over a year later I would return to Green Chimneys with a recovering redtail who needed some time in a large flight cage before release. "How's the peregrine?" I asked.

"He's fine," said Paul. "Stop by and say hello to him before you go."

After the redtail had been released into the big flight cage I wandered down the bamboo-edged path to a cluster of roomy raptor enclosures. There was

Pancho, settled comfortably on a high perch. His coloring was a bit lighter, as he was heading into his adult plumage. His beak, still split, was serviceable but far from perfect.

I felt my throat tighten. It wasn't what I had wanted for him. I thought of all the people who had worked so hard to rescue and restore him to health; it wasn't what any of us had wanted for him. Perhaps one day he would be part of a captive breeding program, and his descendants would fly freely over the Hudson River. Pancho himself would have a comfortable life. He would never again know hunger, never again lie injured and alone. But he would never live the way he was born to live, with nothing but the sky before him.

I looked in. He glanced at me impassively, then looked away. It won't matter if he doesn't remember me, I told myself. It all happened a long time ago.

"*¡Hola, Panchito!*" I said softly. "*¿Cómo está mi pájaro guapo? ¿Te sientes mejor?*"

His head snapped back in my direction, and slowly he cocked his head to the side. Anything good on the menu?

Cuídate mucho, mi pájaro guapo. Que tu vida sea llena de cosas buenas.

Take care, my handsome bird. May your life be filled with good things.

Chapter 24
SOLIDARITY

I bought a programmable space heater, a long table, and a set of plastic drawers. The shed was drywalled and painted. John installed a fluorescent ceiling light and a spotlight over the exam table, which was an old card table I had dug out of the basement. The kids helped me move in all the crates and equipment I had accumulated. I made a sign:

BRING THEM BACK, THEN LET THEM GO

and we hung it over the doorway.

Occasionally I would refer to it grandiosely as "the Clinic," but mostly it was "the shed." And once it was set up, I was ready to go to my first New York State Wildlife Rehabilitation Council Conference.

Wildlife rehabilitators have a surprising number of groups they can join: the National Wildlife Rehabilitators Association, the International Wildlife Rehabilitation Council, and the International Association of Avian Trainers and Educators, as well as various state and local associations. Most host yearly multiday conferences where rehabbers can attend lectures, participate in labs, go on field trips, buy all kinds of supplies, and schmooze with kindred spirits. At its yearly conference, the New York State Wildlife Rehabilitation Council (NYSWRC, pronounced "nicework") honors a veterinarian for his or her outstanding work with wildlife. This year it would be Wendy, who had been nominated by Maggie and Joanne.

The conference was in upstate New York, the award ceremony the first night. During the first day I attended lectures on avian antibiotics, the reintroduction of bluebirds, wound management, caging for injured accipiters, and a lab on principles of fluid therapy. I wandered through the sales area and bought the basic medical supplies I was missing, all the while thinking that I'd better get cracking on my money-making newsletter. I found Denise, who had sent the heron to me, as well as several other rehabbers I had met through the years. I chatted with rehabbers I'd never met, all of whom were cheerful and friendly and welcomed me as one of their own.

When the dinner hour arrived I entered the dining room and quickly spotted Maggie, Joanne, Wendy, and Wendy's husband, biologist Fred Koontz. As I made my way toward them I passed a table where five or six rehabbers were eating heartily, all listening to a woman describe a raccoon who had been caught in a trap.

"This deep," she said, putting down her buttered roll and holding her fingers two inches apart. "The tendon's in ribbons. I go poking around inside, and the whole thing is filled with maggots."

Instead of recoiling in horror, her dining companions leaned closer. "Was it necrotic?" demanded a man sitting nearby, holding his lasagna-laden fork in midair. "Any gangrene?"

I felt a quick rush of exhilaration: *it was dinner time, yet my conversational topics were limitless.*

I didn't know the four other people at Wendy's table. I took the empty chair between Maggie and a friendly-looking woman, and introduced myself.

"Jean Soprano," the woman replied, shaking my hand. "From Pennellville."

"What do you do?" I asked her.

"Bears," she said.

"Get out!" I said, before I could stop myself.

"I do!" she said. "All the large carnivores—bears, coyotes, bobcats, foxes, raptors."

"You have to come visit me," I said, "so you can tell my husband how good he has it."

"Get Jean to tell you a bear story," said a man sitting across the table.

"Please!" I said to her. "Tell me a bear story!"

"Come on, Jean!" urged the other three.

"Okay, okay," said Jean good-naturedly. "Here's a good one. This just happened, so none of you have heard it. So, one night about ten o'clock a couple is driving down a dark road in one of those little hatchbacks. They come around a corner and right in the middle of the road is a big female black bear. The guy slams on his brakes, skids down the road, and bang! Plows right into the bear. Knocks her cold."

"Ahh! Arghh! Oh, no! Poor thing! How badly was she hurt?" the whole table chorused. When it comes to stories involving cars and wildlife, there is no better audience than a group of rehabbers.

"The couple both jump out of their car and run over to the bear, who is lying unconscious in the middle of the road," continued Jean. "They were really upset, and they didn't have a cell phone, and they didn't know what to do, so somehow—and don't ask me how they did this, she was a big bear—they pick her up and put her in the back of their little tiny car."

"They put a *bear* into the *back* of their *car*?" we all gasped.

"Then they drove like hell home. They get home and the bear's still out, so they back the car up to the garage, haul her out of the car and onto a blanket, then they shut the garage door and run inside, where they proceed to call everyone they know. Everyone either asks if they're crazy or hangs up on them.

"Finally they call the dog warden, who says she'll be right there. The dog warden arrives—all five foot two inches, ninety pounds of her—and immediately takes charge. All three of them go back into the garage, where the bear's still out, and they pick her up and put her into the dogcatcher truck, and the bear and the dog warden go off.

"The dog warden brings her to a local rehabber, and the two of them get the bear—who's still out—into a big crate and latch the door. The rehabber calls me, I arrive the next morning, and we put the crate, with the bear inside it, into the back of my car. At this point the bear is conscious but still groggy, so I head off to my vet, John Davis, who works with another vet, Kevin Hammerschmidt. As we're all dragging the crate into an exam room John discovers that the latch is broken, which means that at any time during the hour's drive to their office she could have gotten out and I would have had a groggy bear loose in my car, just like that couple in the beginning of the story."

The gathered rehabbers were all howling with laughter, wiping tears from their eyes and elbowing each other in the ribs; Jean, laughing herself, tried her best to continue.

"John and Kevin open the door a crack to look inside, and evidently the bear was feeling better, because she comes barreling out and starts trying to dig a hole through the office wall. So they throw on their big gloves and wrestle her down and give her a shot of tranquilizer, then as soon as she conks out again they drag her over to the X-ray machine and find out she has no fractures, just a concussion and probably a whopper of a headache. They put her in the crate and wire it shut, and put the whole thing back in my car, and when I finally drove away I could tell they were really, really happy to see me go. The bear spent three weeks with me and then I released her way up north."

We all clapped, drank more wine, and swapped more stories, and later gave Wendy a standing ovation when she received her award. I returned home two days later bursting with inspiration, my wallet crammed with phone numbers and my notebooks heavy with information. John and the kids detailed their lives since I had been away, including sporadic sightings of our released birds, appearing occasionally for food but all of whom were spending more and more time away from the yard.

We saw Null and Void more than the others, although they, too, would disappear for days at a time. Then one Sunday afternoon Mac and I were clearing barberry bushes from an area halfway down the driveway when we heard a strange sound. It was loud and grating, as if a heavy metal object was being dragged down a concrete road. Mac and I glanced around, then at each other. "Grackle migration!" he said.

As we hurried up the driveway the huge flock of grackles flew over our heads and blanketed the woods around the house, perching on tree limbs, poking through the fallen leaves, and emptying the bird feeders, all chattering noisily and turning the land dark with their sheer numbers. Skye hurried out of the house and John out of his office, and we all sat on the deck to watch the spectacle. A half hour later they began to disperse. For the next two days the huge group would come through once or twice a day, eat the seed we'd thrown on the ground, and take off again. One day they failed to appear, signaling that they had moved on and that we wouldn't see them again until the following fall.

For a month I continued to put out the usual plate of grackle food. It would be eaten by an assortment of passing birds or, if it wasn't gone by nightfall, finished off by an opossum or coyote. But our swaggering yellow-eyed bandits were gone.

"They joined the other grackles," said Skye firmly. "They have a family now."

"A really huge family," said Mac. "Think about it—it would be like having a thousand Skyes around you all the time."

"Shut up, Mac!" said Skye.

Putting on my sweats I ran to where the woods were thick with sugar maples. The late October sun poured through their brilliant yellow leaves, turning the forest into a glorious, glowing cathedral. I sat on a fallen tree deep in the radiant woods, as I did every year, listening to the birdcalls and running my hands through the crisp autumn air. Believing, just for a moment, that the world was a perfect place.

Chapter 25

THE THANKSGIVING GUEST

"White or dark meat?" asked my brother Skip, who was hosting that year's Thanksgiving dinner.

"Uh . . ." I said, hoping no one would notice the question. "I'll have a drumstick," I whispered.

"Why are we whispering?" he hissed loudly.

"*I know why!*" trumpeted Skye from across the room. "*You're not eating turkey, are you? How could you?*"

"That's nice," said Mac darkly. "Now you can go home and tell Gravy you ate one of her cousins."

"But I'm not actually eating Gravy," I said.

"Why can't you eat gravy?" asked Skip, who had grown used to my various obsessions. "Because of all the little flour particles that had to die?"

"Gravy is a wild turkey," John explained. "She's home in our clinic."

"Mom named her Gravy," said Mac, "even though I said it was inappropriate."

"I named her Wavy Gravy," I said. "I keep telling you, if you're offended by Gravy call her Wavy. Personally," I said to Skip, "I just call her The Turkey."

"Wavy Gravy was before your time," said John to my brother. "He was a counterculture hero of the sixties."

"And now you've named a wild turkey after him," said Skip. "Should I dare to ask why?"

"You don't want to know," I said.

The turkey in question had arrived a few weeks earlier, after Bonnie and Gary Van Asselt, two local birders, had watched her fall to the ground after she attempted to fly to a low branch in their yard. Searching for help, they found me through the birding grapevine. I told them to bring her over and raced off to my purple three-ring binder to look up "wild turkeys," then hurried out to the shed to set up an area for her.

I had spent the last few weeks gathering an impressive assortment of donated items: waterproof tarps, more crates, another table, old blankets and towels, a large egg-crate foam mattress, bowls, pans, a long foldable dog pen, and various bird food items—including a large half-bag of turkey pellets. I covered one corner of the shed floor with a tarp and several layers of newspaper, then cut a rectanglar section of the egg-crate mattress and covered it with a towel. The turkey would be a double pioneer: my first wild turkey, and the first bird in my shed.

Bonnie and Gary arrived with a large cardboard box. When I lifted the turkey out she staggered, favored her left leg, and seemed slightly dazed. I wrapped her in a towel, told her rescuers they could call back in a few days to check on her, and carried her out to her hospital room. When I put her on the floor she lurched over to her towel-covered bed and sank down, regarding her surroundings with half-lidded eyes, as if it were all she could do to stay awake.

She was thin but had no wounds or broken bones, nor could I see any signs of bruising. She did seem especially warm to the touch, and I could hear a slight click in her lungs as she breathed. I gave her fluids, and an hour later she picked at a few turkey pellets soaked in warm water. Normally I never touch an adult wild bird if I can help it, especially new ones. But for some reason as I was getting ready to leave I reached out and touched her gently on the side of her face. In response she slowly lowered her head until it rested on my upturned palm, then closed her eyes.

"Bonnie!" I said through the telephone, soon afterward. "Was that a tame turkey you brought me this afternoon?"

"No," said Bonnie. "She's part of a big flock that comes to our feeder every afternoon, but they're not friendly. I couldn't get near any of them, until today."

When I called the Croton Animal Hospital the following day I discovered it was Dr. Popolow's day off. "But Dr. Hoskins is here," said Charlene Congello, the receptionist, helpfully. "Maybe he can see your turkey."

Soon I was on my way to the hospital, the turkey riding in a large dog crate in the back of the car. As I pulled into the parking lot I considered asking one of the technicians to help me carry the entire crate into the office. But since everyone at the hospital—the vets, receptionists, technicians, and animal handlers—were dealing with these wild birds and me out of the goodness of their hearts, I didn't want to push it. I stuck my head in the office door, greeted Charlene, made sure the coast was clear, and returned to the car. Wrapping the turkey in a large towel, I carried her into an examination room.

Despite her condition she was a striking bird. Her feathers, beautifully patterned in white, brown, and black, had a dazzling metallic sheen and her head, wrinkled and wattled, was a delicate watercolor of red and blue. She lay quietly, surveying the room with huge brown eyes. She looked kindly but startled, like someone's elderly aunt who, after several days of feeling poorly, had suddenly found herself dumped into the middle of a busy casino.

In a few minutes a tall man wearing a lab coat entered the room. "Bruce Hoskins," he said, extending his hand.

Working with Dr. Hoskins is the closest I will ever get to vet school. Whenever he examines one of my birds he gives me an impromptu tutorial on the injury, the system involved, the surrounding skeletal structure, and the pros and cons of various drugs, until my head feels like it will explode with information. Few experiences can be more terrifying to a wild bird than a veterinary exam, but Bruce's gentle manner always seems to make a bad situation better.

Bruce listened to me recount her history, looked her over, and then took her into the back for an X-ray. The X-ray showed no lead—always a possibility—but did show a cloudiness in her lungs. "What do you think?" I asked.

"Hmm," he said gravely, looking up from his notes. "I'd say 350 degrees for four hours."

I burst out laughing. "I'm sorry," I said to the turkey, wiping my eyes. "But you have to admit you're kind of a comical figure."

Bruce said it could be pneumonia and prescribed a round of antibiotics, adding that if she didn't improve he would do blood tests. I gathered up the turkey and headed home, hoping that antibiotics, food, and rest would do the trick.

She spent the next three days either resting or sleeping on her towel-covered bed. For two days she wouldn't eat, so I had to syringe liquid food through a long rubber tube down her throat and into her crop. On the third day she started picking at the food I'd left in her pen. The following day I opened the door and found that a good portion of her food had disappeared and she was standing and preening, a clear indication that she was feeling better.

By coincidence, one afternoon I received one of those "nuisance wildlife" calls, a few dozen of which I had fielded during the summer. Usually the caller, who has just moved into a brand-new subdivision ripped and blasted out of a formerly beautiful woodland, is exasperated to find that there are actually wild animals in the vicinity and wants to know how to get rid of them. I learned at the conference that "I'm sorry, but I've already promised the animals that I'd help them get rid of *you*," is not the correct response, and that I should seize the opportunity for education. I try, but some callers are more difficult than others.

"My name is Dr. B——," said the male voice. "And I'm having a terrible problem with a wild turkey. I live in a very exclusive area and I paid a great deal of money for our home. This turkey is injured, he can't put any weight on his leg, and he keeps hanging around and doing his . . . his . . . business on my imported limestone. Can you come and get rid of him?"

"Uh-oh," said Skye, looking up from her homework and seeing my expression.

"You'll have to catch him," I said. "Throw a blanket over him, then pick him up and put him in your garage. I live an hour away from you, so I don't want to drive down and find out he's not there."

"Oh, no!" said Dr. B——. "I'm not going to touch him!"

"Look," I said, "try to work with me here. If he can't put any weight on his leg it's probably broken, so he won't be hard to catch."

"Broken!" said the astonished Dr. B——. "Turkeys have bones in their legs?"

"Do turkeys have bones in their legs?" I repeated. "What kind of a doctor are you?"

"An orthopedic surgeon," he replied.

Meanwhile, our turkey was feeling better. I debated putting her in the flight cage. On one hand, the clicking in her lungs was gone and she needed more space; on the other, she hadn't gained any weight and the October nights were chilly. I ended up putting her into the more natural environment of the flight cage. I was quite pleased with my decision until the following morning, when I went to feed her and discovered a nasty wet rasp emanating from her chest. I described it over the phone to Bruce, and soon the turkey and I were back in his office.

I had chalked up her original tame manner to sickness and weakness, but she remained docile and easy to work with despite her improvement in health. Bruce entered the room, looked at the turkey standing matter-of-factly on the table and me leaning casually against the wall, and raised his eyebrows; in return I gave him an elaborate shrug, and we went on from there. He listened to the turkey's loud, rattling breathing, then anesthetized her for a tracheal wash, in which a tiny bit of liquid is syringed directly into the lungs, sucked back out again, and analyzed. Afterward he took another X-ray, then held it up to the light and pointed to her lungs.

"This is a different view, but you can still see the cloudiness," he said. "I took some blood, which I'll also send out. But if she's also not gaining weight it's probably aspergillosis."

Aspergillosis is a respiratory disease caused by fungus. Many wild birds harbor the fungus, which does them no harm unless they become sick or stressed. I moved the turkey back into the shed, where it was warm and dry. Since there were no other birds in residence, at night she slept in a large crate and during the day she was free to move around. She either walked about the floor or hopped up on the table next to the window, where I'd see her small head peering out at me when I approached.

"It's asper," said Bruce, calling later that week with the test results. "I'd like to put her on ciprofloxacin as well as itraconazole, but these drugs are expensive. I'll call the prescriptions into one of our pharmacies, then you can call and find out how much we're talking about."

I dialed the number, apprehensively wondering how much it could be. What if it was . . . say . . . a hundred dollars?

"A month's worth of the two prescriptions . . ." said the pharmacist. "Let's see; that would come out to about eight hundred dollars."

"What?" I sputtered. "Are you kidding?"

"I'm afraid not," he said. "And if you do give the turkey the drugs, you have to check with the Food & Drug Administration to see how long you have to keep her off them before you let her go."

"Why?" I said.

"Because if you let her go and somebody shoots her and eats her, they could be affected by the drugs she's been on."

"Get out," I said weakly.

"I'm going to make a file, just in case you decide to do it," said the pharmacist. "Does the turkey have a name?"

I sat for a moment, trying to come up with a name but unable to ignore the absurdity of the larger picture. I had just started putting together the paperwork for my nonprofit corporation, so I had yet to take in any donations. Eight hundred dollars for one bird was not in the cards—especially a turkey who, if all went well, would be released in the middle of turkey hunting season. I tried to fend off an attack of existential angst by concentrating on coming up with a name.

Logically, of course, I didn't have to. But damn it, I was going to solve at least one part of this problem, and I was going to do it right then and there.

Unfortunately, my brain was not cooperating. "Lassie," I thought. "Mr. Ed." I contemplated some of the more florid names bestowed by sentimental rehabbers, and wondered if the pharmacist could fit "Most Precious and Beloved But Ultimately Unaffordable Gift from the Heavens Above" onto a little pill bottle sticker.

"Are you still there?" asked the pharmacist.

"Her name is Gravy," I said.

"Excuse me?"

"No—it's Drumstick," I said, and immediately felt a pang of remorse. I returned to the less offensive Gravy, which a furious bout of free-association turned into Wavy Gravy. Perfect, I thought. Who better to understand the absurdity of my predicament—and life in general—than the man who said, "I am an activist clown and a frozen dessert"?

"Her name is Wavy Gravy," I stated firmly. I was rewarded by silence.

After hanging up the phone I sat at the kitchen table, trying to figure out how I could come up with the $800. With the drugs she would live, without them she would die, and I was in too deep to call the whole thing off. I called around to see if I could find a cheaper drug source, and discovered that one of the reasons ciprofloxacin was so expensive was because it was being stockpiled by wealthy New Yorkers in case of a post–9/11 anthrax attack. It didn't take long for me to work myself up into a state of white-hot indignation.

"Hi," said John, innocently entering the room.

"Damn these selfish yuppies!" I shouted in response. "Hoarding all the cipro when I have a turkey who really needs it! What are the odds of an actual anthrax attack, anyway?"

"Gotta go," said John, heading back out the door.

"No, wait!" I called after him. "Come back! Listen, that turkey has a fungal infection and I need $800 to fix her."

"What?" he said, looking at me as if I'd suddenly lapsed into an African click language. "Now I've really gotta go," he said, and disappeared.

"And they say rehabbing is all fun and games," I groused.

Chapter 26

STRANGE ALLIANCES

"Here," said Mac, handing me a small wad of one-dollar bills. "You can have it for the turkey."

"What a great kid you are," I said, giving him a hug. "Thank you. But let me see what I can come up with before I take your money."

I went to my desk and started shuffling through the piles of books and papers, searching for my Willowbrook Pharmaceutical Index and hoping it might have a hidden chapter on where to get cheap drugs. Coming up empty-handed, I pulled out my blue three-ring binder entitled "DRUGS" and riffled through it: "Avian antibiotics, from A to Z." "Corticosteroids, use of. Dosage charts." "Pain medication." "Worming medicine." Nothing. I picked up the notebook I'd taken to the conference. Its pages were covered with my ragged script, its pockets crammed with lecture handouts and flyers from various rehab centers. I skimmed through my notes: "Effects of lead poisoning." "Treating bumblefoot." "Songbirds and homeopathy." "Basics of parasitology." Just as I started wondering if the turkey and I should head for Canada, something caught my eye.

call Erica Miller DVM T St BR for chp itcnzle
AZ 1/10 cost, will overnight

I flashed back to a large lecture room where Dr. Erica Miller, staff veterinarian of the reknowned Tri-State Bird Rescue in Delaware, was giving her talk on avian antibiotics. Drugs can be prohibitively expensive, she had said, but there was a pharmacy in Arizona that sold compounded (mixed into a liquid or gel) medication for a tenth of the regular price. I lunged for the phone, called Tri-State, and scribbled down the number. Filled with sudden hope, I dialed Pet Health Pharmacy.

"Cipro and itraconazole," said the cheerful woman who took my call. "A month's worth of the two prescriptions. Let's see . . . that would come out to about eighty dollars."

"No!" I said, awestruck.

"And did you say you were a rehabber?" she asked. "Because we give discounts to rehabbers."

"Life just keeps getting better," I said.

Not everyone was impressed with my discount.

"Eighty dollars?" said a doctor acquaintance. "That's still a lot of money when there's a chance somebody's going to shoot her."

"Don't give me that," I replied belligerently. "Say you're working in a clinic in a dangerous neighborhood and I come to you with pneumonia. Are you going to refuse to treat me because there's a chance I might get mugged on my way home?"

"No," he replied. "I'm going to refuse to treat you until you're wearing a proper straitjacket, like you should be right now."

Three days after the medicines arrived the turkey's loud wheezing and coughing were gone, and from then on it was just a question of keeping her somewhat calm until her lungs healed. When I put her in the flight cage she paced endlessly back and forth, wearing a trench in the dirt, instead of lounging around gaining weight as I'd requested. Most days I let her loose in the shed. If she felt like it, she could move around on the floor, where she had no view of the woods. If she wanted a view, she could hop up on the table, but she had to stand still in order to look out the window. This worked relatively well, until the turkey vulture arrived.

As always, it began with a phone call. "It's a great big vulture," a man's voice said indignantly, "and somebody shot him."

Rehabilitators dread the start of hunting season. On the one hand, you have the conscientious hunters who know the woods, eat what they hunt, and wouldn't dream of shooting a protected species. On the other, you have the drunken idiots who are just out there to blast whatever has the misfortune to cross their path. The turkey vulture I eventually found huddled next to an old house in Westchester had obviously met up with one of the latter.

Turkey vultures, along with their smaller cousins, the black vultures, are smart, funny, and sociable creatures. As they soar through the air they rock gently back and forth, like huge black butterflies, and they greet a bright morn-

ing by perching together and spreading their wings like an ancient clan of sun worshipers. I find large groups of circling vultures so exhilarating that I become a dangerous driver; when my kids are in the car with me, they respond to the sight of vultures with howls of "Watch the road!"

"The jury's out on this one," said Wendy, after X-raying the Westchester vulture, removing several shotgun pellets, and thoroughly cleaning and setting his wing. "The bone is splintered, so we'll have to see how it heals. There might be nerve or tendon damage. But it's worth a try."

Turkey vultures look like wild turkeys designed by Tim Burton. Both are large birds with unfeathered, wrinkled heads. But turkey vultures, so named because of their resemblance to wild turkeys, strike the average person as more sinister because of their black plumage, smaller eyes, ivory-colored hook on their beaks, and the superstitious nonsense that has dogged them ever since humans started making it up. Sure, vultures eat dead things; the last time I looked, so did *Homo sapiens*.

I put the turkey in her large crate on one side of the shed and the vulture in another large crate on the other, so each had a window. The settling-in period for vultures can be difficult. Some are immediately matter-of-fact about dealing with humans; others are horrified by the very idea. This particular vulture fell into the latter category. For the first two days he would greet me like a drunken frat boy caught by the college dean: he'd eye me with alarm, empty the contents of his stomach, then regard me reproachfully, as if the whole sad state of affairs was beyond his control and entirely my fault. Except for my twice-daily hospital rounds, I stayed away from the shed. Yet in spite of my lurid descriptions, the new patient had two determined visitors.

"Awww, come on, please let us see him," begged Skye. "Pleeeeeeeeeze!"

"We'll be so calm and quiet and we won't upset him at all, we promise!" said Mac.

"Don't say I didn't warn you," I said, leading them into the shed, where the vulture took one look at us and heaved up a good-sized pile of partly digested venison. We all turned, closed the door, and trudged silently back to the house. "Can you drive me to Cindy's house?" asked Skye. "*Her* mom is a hairdresser."

By the third day the vulture had decided that I was a relatively benign figure who brought food and didn't hurt him, and from then on he would vomit only when he was forced to undergo something he considered truly ghastly, like a bandage change. For the daily cleanup I would open the crate door, step to the side, and poke a thin stick through the back panel; the vulture would step away from the stick, out of the crate, and into a small pen. We'd studiously ignore each other while I shuffled bowls and newspaper, and during those few occasions that I caught his eye I'd immediately drop my gaze, thus proving that I was far more scared of him than he was of me. Since a large percentage of injured wild birds eventually die from the stress of being in captivity, the vulture's growing comfort level was a milestone in his recovery.

I roped off a third of the shed and hung a large sheet as a privacy curtain, so the turkey could come out of her crate and have some space to move around. Eventually she finished her medication and I began putting her in the flight cage during the day, when it was warmer, and bringing her in at night so she could continue to gain weight.

After two weeks Wendy took another X-ray of the vulture's wing. The X-ray showed calcium forming around the shards of his splintered bone but not enough to leave the wing without support, so she wrapped it for another week. He was doing well, though, and needed space, so I put him in the left side of the flight cage and set up a heat lamp in a corner where he could go to warm up. He was obviously happy to be out of the crate. He walked around, tested out the low perches and logs, stretched his good wing, and continued to eat heartily.

But the status quo never stays so for long. "Hey, Suzie?" said Joanne through the phone. "Can you overwinter a couple of pigeons in your flight?"

If a bird is injured in the autumn and recovers, it often means the rehabber will still have to "overwinter" him, or keep him until spring. A migratory bird normally in Costa Rica for Christmas can't be booted out into the snow and be expected to survive. Even birds who spend the winter in harsh climates are sometimes kept through the winter, as letting them go in the spring will give them a better chance. This means that there are many healthy birds biding their time until spring, and many rehabbers wishing they didn't have to spend the winter cleaning up after them.

"I guess," I said. "Who do you think they'd rather have as a roommate—a wild turkey or a turkey vulture?"

"They don't care," said Joanne. "They're pigeons."

I posed the question to the family. "Put them in with Wavy Gravy," said Skye. "Because the vulture might get nervous and throw up on them."

"But the turkey is leaving soon," said Mac. "Put them in with the vulture— he needs some company."

"Put one in with the turkey and one with the vulture," said John.

"How about if I put them both in the bathroom?" I said.

"Don't even joke about that," said John.

Mac had a point: the turkey would be leaving soon. The vulture was well-fed, had plenty of space, and vultures are not an aggressive species. The pigeons could fly while the vulture couldn't. And, as any soldier will confirm, odd alliances can be formed when you're in a captive situation. Why not give them a chance? I propped their crate door open, and the pigeons walked busily into the flight cage.

Pigeons look surprised even if there's nothing going on, so it was a little hard to tell just how undone they were by the sight of a large turkey vulture in close proximity. After a single frozen moment they flew up onto a high perch, where they sat staring at the strange creature below them. The vulture stared back with his usual look of wary interest, but didn't seem particularly impressed.

As the days went by, the pigeons became more intrigued by their flightmate. Despite their bad publicity, pigeons are curious and intelligent birds. They are successful and adaptable without being aggressive toward other species; they have complicated and interesting relationships; and they simply love to fly, as anyone who has ever seen a flock wheeling through the sky can attest. The two in the flight cage seemed irresistibly drawn to the mysterious vulture: the first day they stayed high above his head; the next day they perched just slightly above him; and by the third day all three were sitting on their own tree stumps, each no more than a few feet apart. I'd peer into the flight cage and see the vulture walk regally past me, followed by the pigeons, like a rock star trailed by his fans; unable to stop ourselves, we dubbed them Jerry Garcia and the Deadheads.

Jerry's shattered bone healed slowly, and there was permanent soft tissue damage. While the wing was immobilized, the injured muscles and tendons contracted further, eventually preventing him from extending it more than halfway. I'd appear each morning (when his stomach was relatively empty), toss a small towel over his head to calm him down, then sit on a log and slowly and carefully extend and retract the wing, hoping the tendons would stretch and regain some of their elasticity. No one appreciated my efforts, of course; the Deadheads would fly to a high perch and watch me with suspicion, and every once in a while Jerry would snake his beak out from under the towel and bite my hand. Although vulture bites can be quite painful, they are not in the same league as, say, pet parrots, who can practically sever your finger when they're in a bad mood.

Eventually it became clear that Jerry's soaring days were over, and I found him a home at a sanctuary that was looking for a companion for their single vulture. I tried to include the Deadheads in the deal, but they turned me down. The flight cage seemed empty without Jerry, who would eventually forge a companionable bond with the sanctuary's vulture. For days the pigeons wandered forlornly around the flight, two little groupies deprived of their rock star. All they really needed was the company of their own kind, and before long their own kind began to arrive. In spades.

But meanwhile, Wavy Gravy had became more and more rambunctious, making it clear that she considered it time to go. When her release date finally arrived, I called Bonnie.

"Your turkey's all set," I said. "I didn't think this day would ever come. I want to release her back into her flock—what time do they show up?"

"They don't," said Bonnie. "They've moved on. I haven't seen them for weeks."

I called everyone I knew in Bonnie's area; no one had seen the turkeys. I called my other birding friends, thinking a strange flock would be better than no flock at all, and fared no better.

"I don't know what I'm going to do with her," I said to my friend Diana Swinburne. "I can't find her a flock, and I don't want to release her completely alone."

"Why don't you take her to Anna's?" said Diana. "She could join the turkey parade."

Anna Schindler lives in a house overlooking the Hudson River and has a soft spot for wild turkeys. They disappear during the spring and summer, then appear faithfully at her door throughout the fall and winter. Anna feeds them generous helpings of corn, and in return they accompany her on her daily trek to the mailbox. She walks down her dirt driveway, followed by fifteen to twenty turkeys; when she reaches her mailbox, she turns and orders them to stay away from the road. They stop where they are and mill about while Anna looks over her mail, then they step aside for her and follow her back to the house. I called Anna and introduced myself. Would it be all right if I released a wild turkey at your place? I asked her.

"Oh, sure," she said in her broad German accent. "What's one more turkey?"

Anna called me the next morning at 8:00. "They're here," she said.

There were fifteen wild turkeys on Anna's front lawn when I arrived. Our turkey hopped out of her crate sporting one bright orange leg, colored that morning with livestock crayon for temporary identification. Just as she was getting her bearings, another female barreled out of the group and launched herself at the intruder. It was here that Wavy Gravy the turkey parted philosophical company with her namesake; this turkey was no pacifist. She lit into her attacker, furiously pecking her, biting the back of her neck and whacking her with her wings; when a big male strode over to join the fight, she clobbered him, too. Finally she chased them both away, left the group, and walked around in large exploratory circles. Anna called me that afternoon to report that the turkey with the bright orange leg was peacefully scratching at the ground with the flock. By the second day there was barely a trace of orange left, and by the third it had disappeared all together, along with my only way of staying in contact with her. For the next few weeks I felt a stab of fear each time I heard the echo of a shotgun. I'll never know if she stayed with her new flock or eventually left in search of her family, but at least I know she did it with a clean pair of lungs.

Chapter 27

PEYTON (PIGEON) PLACE

The snow arrived in time for Winter Solstice. It was a classic New England snowfall, too peaceful to be called a storm, falling in big, heavy flakes and blanketing the woods with six inches of powder. The kids raced outside, pelting each other with snowballs and rolling down the hills. A few days before I had tarped the roof of the flight cage, balancing precariously on the narrow beams as I unfolded the heavy blue plastic and used short nails to secure it. I covered the loft under the roof of the second flight with hay, giving the two pigeons a snug spot in which to ride out the winter.

I had no other wild birds and was grateful for the break. During Christmas vacation we went sledding and skating, took moonwalks, and played Monopoly in front of the fireplace. We spent hours working on old Springbok circular puzzles, unable to pull away until the final butterfly or garden flower was complete. John and I caught up with friends, few of whom I'd seen over the summer.

"Alan thinks you're ignoring him," said Jan, Alan's veterinarian wife, when they were over for dinner one night.

"Alan," I said. "You have to do me a favor—move your office down here. It's a ninety-minute round trip to see you, and there just aren't enough hours in the day."

"No problem," said Alan. "I'll get right on it."

When the kids went back to school I filled out my nonprofit application, e-mailing the saintly Bob Bickford with so many questions that I was sure he had to regret his offer to help me. But he never failed to provide quick and sure answers, and the recurring spasms of terror I felt about inadvertently crossing the Internal Revenue Service were always allayed by his knowledge and confidence. Finally all my paperwork was finished, mailed, and on file. Flyaway, Inc. would receive its nonprofit status by fall, and I could start soliciting donations.

Randi Schlesinger took the text of my newsletter and, through her graphic designer wizardry, created a work of art. She trudged through the snow into the woods, pointed her digital camera straight up, and took a photograph of the latticework of winter branches silhouetted against the sky. Manipulating it until it was faded and dreamlike, she extended the image into an elongated rectangle, then, using an airy, graceful font, dropped *flyaway* on top of it. The result was an eye-catching, memorable masthead, and an image—of branches just waiting for a bird to land on them—that could be used as the background for headlines throughout the publication.

The four-page foldable newsletter had a section for basic information, one for thank-yous, a panel for donations, and a large middle area for bird stories. By the end of the year I had taken in eighty-two injured or orphaned birds, so I had quite a few from which to choose. I selected four feel-good stories and wrote them up. "I love these stories!" said Randi, her infectious enthusiasm making me giddy with anticipation. "Do you have pictures? I'll put a photo of the bird by each story, and then two or three more right next to the donation panel. This is great! I *know* people will send you money!"

And she was right. I sent out about fifty newsletters and friends, relatives, and members of our tight-knit little community responded quickly and with generosity. Touched and gratified, I wrote thank-you notes and deposited the checks into my official Flyaway bank account. I felt dangerously on top of things. Eighty-two birds between May and December made for a hectic summer, but it was certainly doable and didn't seem to be causing my family any distress, apart from the occasional disgruntlement of a missed movie. I had a nice break during the winter, when I could relax with my family, write my newsletter, and coordinate the growing treasure trove of avian information stored in my rainbow-colored library of three-ring binders.

I had no idea what was to come.

By mid-March there was a flurry of pigeon activity. "It's Joanne again," she said. "Say, can I unload two more pigeons on you?"

"I don't mind," I replied. "I'll put them in with the Deadheads."

"The dead what?" said Joanne, with alarm.

There is something inherently comical about pigeons. Maybe it's the way they bob their necks back and forth as they walk briskly about, as if they're on a very tight schedule and should really be somewhere else. Maybe it's that Three Stooges *whoo-whoo-whoo* sound they make when they take off. Maybe it's the fact that once they're comfortable around you, they don't seem to care what you see them doing. On a few occasions I've even suspected them of overacting for their audience, like reality show contestants, although this is not something I could convince any self-respecting biologist to believe.

Joanne arrived with the pigeons: Sly (charcoal gray, broken wing healed but slightly stiff) and Sasha (light gray, hit a window but recovered). I don't name all the birds who end up here. Some don't stay long enough, and for some no name seems appropriate. Some are so wild and resentful of their confinement that to name them would feel like harnessing them with another symbol of captivity. But this eventual gang of pigeons were ripe for naming, especially when the group dynamic emerged. Sly was named for action hero Sylvester Stallone, and Sasha for his first wife. The reason will become clear in a minute.

Sly and Sasha, a comfortably bonded pair, spent a few days settling into the flight cage. Initially the Deadheads watched them from a high perch, as they had done with Jerry Garcia. Within a few days all was amicable, although the Deadheads seemed to prefer to observe the newcomers rather than to interact with them. Then came the phone call.

"Aha!" caroled another rehabber friend in an I've-got-you-where-I-want-you tone. "I hear you have pigeons in your flight cage!"

In two days I had three more pigeons. There was BP (Brown Pigeon, healed broken wing, one foot slightly turned inward) and Who Me (young and light gray, caught by a cat and waiting for tail feathers to grow back). And finally there was the avian version of plutonium: Anna Nicole Smith (found emaciated and unable to fly).

Anna Nicole was so huge and blindingly white that she seemed like a different species, and when I released her into the flight cage the other pigeons stared at her with what must have been astonishment. Named for the voluptuous blonde one-time Playmate who "fell in love" with a wheelchair-bound eighty-nine-year-old Texan (who happened to have a few hundred million in the bank), the avian Anna Nicole flew up to a hanging perch and waited for the action to come to her. Which it did, almost immediately.

My celebrity frames of reference tend to date back one or two decades, which was when I began replacing gossip magazines with parasitology textbooks, something of a lateral move. But as I recall, Sylvester Stallone was married to a perfectly nice, normal woman, then he moved to Hollywood and took off like a shot with various gargantuan blondes. Anna Nicole Smith wasn't one of them, but had she been there at the time she probably would have been, so I feel justified in making up this union. Anna Nicole eventually met her untimely end in a Florida hotel room, but at this particular point she was very much alive and busily trying to wrestle her recently deceased husband's estate away from his only child.

It didn't take long for the avian Sly to abandon his devoted Sasha and fly over to Anna Nicole's branch, where he perched in what appeared to be dumbfounded dazzlement. Half the size of the towering Anna Nicole, he looked like an extra from *Zorba the Greek* who'd mistakenly wandered onto the set of *Das Rheingold*. BP and Who Me flew to the ground and started busily looking for seeds, while the Deadheads watched the newcomers intently and Sasha stayed alone on her perch.

"Hound dog!" I hissed at Sly. "Get back where you belong!"

As would become a pattern with the pigeons, he didn't follow my direction.

Luckily for my bleeding heart, Sasha was not the kind to dwell on the past. She spent a day wandering the flight by herself, then the following day I found her cozied up to BP. I was glad that I no longer had to worry about Sasha's mental health until I saw BP chasing Who Me around the flight, pecking his former buddy and pulling feathers out of his neck in a frenzy of newfound machismo. I closed the door between the two flights, separating Sasha, BP, and the Deadheads from Anna Nicole, Sly, and Who Me. Anna Nicole and Sly quickly moved up into the loft and started carrying on like weasels, leaving the young and frightened Who Me ignored and alone. Cursing under my breath, I headed off to the phone.

"What are you doing?" said John, puzzled, when he heard me asking various people if they had a single pigeon. "I thought you didn't want the ones you have."

"I don't," I said darkly. "I realize I'm acting like an idiot, but it's out of my control."

Stripes (light gray with black stripes on lower back, clipped by a car but broken wing healed) arrived two days later. Stripes was a mellow bird who, against all odds, actually did what he was supposed to do: he provided Who Me with badly needed companionship and the Deadheads with new entertainment. Things settled down in the flight cage, just in time for Gladys to bring me another pigeon.

Gladys Rosa lives in a neighboring town and is the wildlife guardian of her small backyard. Resting on her second-floor deck are long trays containing a smorgasbord of seeds and peanuts, frequented by a large population of very happy birds and squirrels. During the previous summer Gladys had somehow spotted a young female house sparrow with her head stuck in a lawn chair; she raced out into a thunderstorm, rescued the sparrow, warmed her up, fed her, then took her to a veterinarian when she couldn't fly. The vet sent the bird to me, and after three days in the flight cage she was good as new. When Gladys appeared she gave me a bright smile and a donation, then returned the sparrow to her rightful place.

"I know you're going to be mad at me," said Gladys sorrowfully, getting out of her car and handing me a cardboard box. "I know the hawks have to eat, and I shouldn't have chased him away. But the pigeon was struggling and the hawk was pulling out all his feathers, and I couldn't help it! I'm sorry!"

I opened the box. Nestled inside a thick towel was a big dark gray pigeon who had obviously neglected to look up when the hawk arrived for lunch. He had a large wound on his back, one under his wing, and one on his head; one eye was closed.

"I'll try to fix him up," I said to Gladys. "But you have to promise me that the next time you see a hawk you'll look the other way. He's just hungry, and if he didn't get this pigeon it means he has to go get another one." I tried to be stern, but the truth is that there's absolutely nothing that Gladys could do to make me really mad at her. Who but Gladys would even notice a sparrow with its head stuck in a lawn chair?

Gladys's pigeon, christened Hawk Food, not only made it through the night but also recovered remarkably quickly. Each day he'd growl and slap me with a wing as I reached in to pick him up, then he'd sit stoically as I gave him his dose of antibiotics, cleaned and treated his wounds, and medicated his eye. His wounds closed, his eye healed, his feathers grew back, and soon he was once again a glossy, healthy pigeon. "Let that be a lesson to you," I told him. "Watch your back."

Winter was over and the flight was like a boardinghouse during Spring Break. I'd walk in to find Sly and BP raucously cooing and spinning, Sasha and Anna Nicole working feverishly on their makeshift nests, and the Deadheads loitering about like determined little Peeping Toms. "Stop that!" I'd say, like a peevish old landlady. I imagined them all snorting and rolling their eyes in response except for Who Me and Stripes, who both seemed to be trying to maintain their dignity in the midst of an unacceptable situation. When I added

Hawk Food to the mix he joined Who Me and Stripes, and for a brief moment I thought we could all hang on until everyone was released. But then: eggs.

After watching Sasha and Anna Nicole sitting solemnly on their nests for days, I climbed up to the loft and found two eggs in Sasha's nest and one in Anna Nicole's. This was not good. Release Day was approaching, and you can't toss a nestful of babies into the air and expect them to fly away. "I'm sorry," I said, reaching under the outraged mothers and taking their eggs. "I hate to do this to you, but you're a wild bird, and as soon as I let you go you can have all the babies you want." Three days later there were more eggs, and the release date was moved up.

Gladys appeared and took Hawk Food, Who Me, and Stripes. She was delighted with her pigeon's recovery, said she hadn't seen his assailant in two weeks, and was happy to release all three in her backyard and provide food for them. I took the rest of the crew to a park on the river, where people feed the pigeons and geese and where I hoped they'd join the existing flock. When I opened their carriers they all sailed off into the springtime air; Anna Nicole landed on top of a gazebo; Sly, Sasha, and BP were in an oak tree; one Deadhead was in a maple and the other right in the middle of a pigeon flock on the beach. I scattered birdseed and watched them for a while, then as I readied myself to go, I glanced up just in time to see a wild pigeon land next to the Deadhead in the maple. Before I had the chance to wonder how they would interact, the two of them were having a hearty party. For this is the way of the wild bird: life is meant to be lived, not just observed.

Chapter 28

TRAVELING FEET

There comes a time in every owlet's life when he decides it's time to hit the road. Complications arise when his feet are ready, but his wings are not.

"Hi, it's Wendy," she said on the phone one April morning. "I have a little owl here, and he's reeeeaaaaally cute."

"How cute?" I said. "And how little?"

"Well," she said, in a philosophical tone. "He's cute and little for a great horned."

Great horned owls are North America's largest species of owl. Although great grays appear to be taller, their added inches are really just feathers; they have the owl equivalent of big hair. Adult great horneds have legendary feather-covered feet that can exert a grip of approximately 250 pounds per square inch on whatever unfortunate creature—or part thereof—they manage to get in their grasp.

Wendy said that the owlet, which at four or five weeks old was too young to fly, had somehow ended up on the ground beneath its nest. The owlet was uninjured, which was remarkable for two reasons: one, because great horneds usually take over crow or hawk nests built near the tops of thirty- to fifty-foot coniferous trees; and two, because soon after it landed it found itself flanked by two large dogs.

Great horned owls lay two to three eggs and they don't hatch at the same time, which means there can be quite a difference in the size of the chicks. The older they get the more jostling goes on, so it stands to reason that somebody occasionally gets elbowed out of the nest, or stands up and loses its balance, or decides to go for an ill-fated stroll. In any case, in a perfect world the chick would tumble to the ground, then hop onto a low branch or a fallen tree, where the parents would continue to feed and protect it. Unfortunately the owl's perfect world has been taken over by humans and their domestic pets, creating a whole new set of problems.

Luckily for this particular chick, the owner called off her large but gentle dogs, put the owlet into a box, and delivered him to the local animal hospital. When I arrived to pick him up, I looked into the box and reacted with the same complex set of emotions that most people experience when they first see a large nestling owl. The creature was incredibly cute—sort of. Fuzzily soft and saucer-eyed, he also sported a black dagger of a beak and stood on huge feet tipped with lethal one-inch-long talons. His intense stare said: "Feed me or I'll kill you." No problem, I said, and took him home to a waiting bag of mice.

The plan was to get the healthy owlet back to his parents as soon as possible, but meanwhile he needed to eat. The last thing you want a young wild owl to associate with food is a human, which precluded my leaning over his box, cooing and making kissy noises as I attempted to feed him. The solution was an owl puppet. By putting my hand all the way up into the puppet and manipulating its beak, I could pick up a small defrosted mouse. And by hiding behind the puppet I could ensure that the mouse was offered, not by a human, but by something recognizably owlish. A scared and hungry baby owl is usually more than willing to take comfort where it can—in this case, in something fuzzy, round-eyed, and holding a good meal by the tail.

Anyone seeing a nestling owl inhale a mouse will quickly realize that owls are the best friends a farmer could have. Owls are rodent-catching machines. They can hear tiny feet rustling through the grass a football field away, and gliding silently on softly fringed wings, will often snatch their prey before it even detects their presence. Young owls are ravenous, especially ones who have

missed a few meals. When my owl puppet dangled a mouse above the great horned nestling he lunged upward, seized the limp mouse by the head, and after a quick series of down-the-hatch motions, swallowed it whole. Three more received the same treatment; after the final mouse, the nestling settled down contentedly, eyes half shut, a long pink tail hanging casually out the side of his beak.

The kids snuck in to see him after he had fallen asleep. Eyes closed, talons covered by downy fluff, he looked like an angelic little cartoon football. "If you promise to do exactly what I say," I whispered to them, "I'll let you feed him next time."

Several hours later I handed Mac a mouse. Mac was born with a pragmatic worldview. He has always matter-of-factly separated the living from the dead, reasoning that once a creature has gone on to its great reward, somebody should make a meal of it. Hiding quietly behind the owl puppet, he held a mouse by the tail and slowly lowered it over the owlet, who snapped it up without hesitation.

"Cool!" he said.

Skye's reaction was more complicated. During the winter she had viewed the burgeoning "raptor section" of the freezer in the garage with growing disapproval, eventually refusing to open the door lest she catch sight of something furry encased in a plastic bag. Although well versed in the intricacies of the food chain, she disagreed with its general principle. Always a champion of the underdog, she would have jumped in front of a Rottweiler to save the little great horned; now, however, the oppressed had become the oppressor. She stopped at the sight of the mouse, who, if truth be told, was quite small and cute.

"Sweetie, I'm sorry," I said preemptively, "but I'm not the one who did him in."

"You didn't do him in," she retorted, "but you're serving him for dinner."

When Skye feels combative, no argument on earth will win her over. I had already covered—unsuccessfully—the balance of nature, the relationship between predator and prey, and the importance of the proper calcium-phosphorus ratio in raptor nutrition. Once I'd even burst into a chorus of "The Circle of Life," which had served only to enrage her. With these recent defeats in mind, I veered off into the afterlife.

"If the mouse has already gone to mouse heaven, do you think he really cares what happens to his body? Maybe he'd be happy knowing he was helping a little owl."

"Mouse heaven?" she repeated, giving me a withering stare. "Oh, *please.*"

Late that afternoon I made a quick trip to the nest site, which turned out to be less than ten minutes from my house. The nest, a large and solid structure, was near the top of a forbiddingly tall hemlock to the side of a dirt road. Just visible above the edge of the nest was the downy gray head of the owlet's sibling, which meant that the parents were somewhere close by. With the exception of nesting northern goshawks, great horned owls are widely acknowledged to be the most ferocious raptors in North America. The problem: how to get a squirmy, uncooperative owlet forty-five feet up a tree without being maimed by its outraged parents. The solution: call Lew Kingsley.

The following morning Lew parked his car behind mine and squinted up at the hemlock tree.

"Mmm-hmm," he said, scowling.

"Hootie!" I trilled, pulling the curtain away from the owl's carrier. "Say hello to your Uncle Lew!"

At the sight of the owl, Lew broke into a wide grin. "Cute little guy," he said. "Nice feet."

Lew tied a weight to a long, thin rope, and without even appearing to aim, tossed it into the air. As if by magic the weight sailed upward, threaded itself neatly around a branch five feet from the nest, and returned to earth. Lew tied the rope around the handle of a bushel basket partially filled with sticks and pine needles, and into the basket went the owlet. The basket would provide a safe and sturdy temporary "nest," and because there were narrow gaps in the bottom, there was no danger of its filling up with water should it rain. When the parents came to feed one sibling, they'd find the other. Few old wives' tales have been more destructive than the one about parent birds refusing to accept nestlings touched by human hands; the parents just want their babies back, and with the exception of vultures, birds have a poor sense of smell—a fact which can be verified by anyone who has watched a great horned owl dig enthusiastically into a skunk dinner.

I couldn't spot either of the parents, although I was sure that at least one of them was no more than fifty yards away, watching intently, probably gauging how much of a threat we were to one chick and mystified by the sudden reappearance of the other. Had Lew tried to climb the tree and replace the missing chick, he might have ended up with one of the parents imbedded in the back of his neck, the possibility of which he was fully aware.

"Not the kind of bird I want gunning for me," he said.

I slipped two mice into the basket next to the well-fed owlet, then Lew hoisted the whole thing up and anchored it firmly. The sibling, who had poked his head over the side of his nest, watched the proceedings with interest. With any luck, the owlet would stay put until he was steady enough to hop out onto a nearby branch, and eventually to fly. That night I returned just before dusk. Standing next to the nest was one of the parents, dark and stately, glowering at me with huge yellow eyes. I could see the chick in the nest as well as the one in the basket. I returned the following day and found a similar scene. Thrilled by the fact that the owlet was still in the basket and not on the ground, I decided to hope for the best and leave them alone.

Unbeknownst to me, however, a few days later a construction worker drove by the nest site and found an owlet on the ground, once again flanked by large dogs. The construction worker was renovating the house of a family who lived a half mile up the road; he picked up the owlet, took it to work, and now, a week later, the owl was still living in the family's barn. The woman who called said she knew they shouldn't keep the owl and hoped any resulting problems could be remedied.

Although it is tempting to keep a baby raptor, it's a terrible idea. Unless you have a federal license, it's illegal; and unless you have a freezer filled with rodents and other assorted raptor food, the baby will develop metabolic bone disease and die. Should you decide to buy some frozen mice, keep the baby for a little while, and then let it go, you are still signing its death warrant: raptors are taught how to hunt by their parents, who continue to supplement them with food while they are honing their skills. A raptor who has no idea how to hunt and who has lost its fear of humans will either die of starvation or be killed when it approaches people for food. A raptor who grows up around children may decide to land on one when breeding season begins. Those whose kids are clamoring for a pet raptor should remember: *The Harry Potter books are works of fiction.*

Fortunately for this owlet, the family temporarily hosting him recognized these potential problems and called me. Quite a bit larger than when I'd last seen him, the owlet was thin and more comfortable around people than he should have been, but was otherwise relatively healthy. The family had named him Boo; on the way home I renamed him Boomerang, since he kept returning to me; by the time he was settled and I had hauled out the mice and owl puppet he was back to Hootie, my generic name for all owls; by the time he had scarfed down four mice in a row, he had no name at all.

"He doesn't get a name," I told John. "He's supposed to be a wild owl. And he's going to be a wild owl if it kills me."

"Right!" said John. "It's back to the woods for the little freeloader!"

If I could just get him back with his owl family fast enough he would forget his time among humans. But first he needed to fatten up, so there were four more days of recorded owl hoots, owl puppets, and endless servings of mice. When my supply of small mice started to run low, I depended on the kindness of friends, who couldn't bring themselves to set traps for the field mice racing around their houses until there was an actual need for it.

"Suzie!" came a whisper on the phone one evening. "It's Kurt Rhoads—I have a whole bag of frozen field mice for you, but I have to bring them over right now. My mother's visiting and if she sees them she'll have a heart attack."

"Here," said Wendy Lindbergh, appearing at my door one afternoon and handing me a gaily wrapped bag.

"Oh!" I said delightedly, picturing a bottle of girly hand lotion or flowery eau de toilette. "What's the occasion?"

"What do you mean, what's the occasion?" she said, as I pulled out a freezer bag filled with mice. "Isn't the owl hungry?"

Normally people feeding raptors do not feed them wildlife, as one can't be sure that the wildlife was completely healthy before it became a food source. But occasionally you can be almost sure, as with these strapping little field mice who had zero chance of having ingested anything toxic. I am lucky enough to live in a place where the majority of people are environmentally aware, wouldn't use poison if their lives depended on it, and are happy to help out when they can.

At dusk on the fourth day I drove to the nest site, trying to figure out how I could once again reunite the footloose owlet with his family. By then he and his sibling were just about ready to start flying, but I couldn't guarantee that they could actually do it—and there were still the dogs to consider. I pulled over and looked up. Gazing back at me was the sibling, who had left the nest and was perched on a nearby branch, and an adult, who was probably the mother and did not appear pleased to see me. My delight immediately turned to apprehension: the two of them could fly away at any time, and my owlet would have no family to return to. I dove back into the car and raced home.

Returning a short time later I climbed out of the car, shrugged into Mac's lacrosse helmet and shoulder pads, and pulled on my elbow-length leather raptor gloves. I reached into a carrier and removed the sizable young owl, who glared and snapped his beak defensively. Walking toward the nest, I held him up over my head. I felt a sudden solidarity with those scantily clad young women who stroll around the ring at boxing matches, holding placards with numbers over their heads; except for the fact that I was middle-aged, wearing body armor, and holding a great horned owl, our jobs seemed strangely similar.

The adult owl froze, then crouched slightly and stared at me with frightening intensity. On the other side of the road, across from the nest, was a heavily wooded area eventually leading to open fields. I started walking backward into the woods, keeping my eyes on her, remembering a friend who took care of an enormous flight cage full of great horneds. As long as he was facing them, he said, the owls would watch him from the highest perch without moving a muscle; but the one time he turned away from them, he received a blow to the back of the head that knocked him flat. He scrambled to his feet to find the owls all in a row, identically impassive, and never found out who had actually committed the assault.

I had no doubt as to the identity of this potential perpetrator. Walking backward through the woods is not an activity I'd recommend, especially while carrying a furiously struggling owlet and being pursued by its outraged mother. I sidled deeper into the woods, trying to avoid tripping over forest debris as the adult flew from limb to limb behind me. Finally I found a huge old leafy maple, probably struck by lightning, that had broken about four feet above the ground. Propped up by the surrounding trees it lay at an angle, its crown a good nine or ten feet high. Cradling the owlet in the crook of my elbow, I climbed onto the trunk and slid up the tree, cursing the fact that I'd been in too much of a hurry to change from shorts to long pants. When I reached the base of the tree's crown I placed the owlet ahead of me, where he had a wide choice of branches and limbs on which to perch. I inched back down the tree, hoping the adult would accept the strips of skin I was leaving on the maple's bark as penance for touching her baby and grant me a few more minutes of clemency.

Glaring venomously she let me slink away, then flew closer to where her owlet perched. For several minutes I watched through the trees, hoping to see the reunion, but as long as I watched she would go no farther. I emerged from the woods dirty, slightly bloody, and trailing various pieces of lacrosse equipment. A passing car slowed down; the occupants gave me a quick once-over, then sped up and drove away.

The matriarch must finally have had enough of human interference. When I returned early the next morning, all three owls were gone.

Chapter 29

SIGNS OF SPRING

It was spring, it was warm, the trees were budding, and I was so happy to see the returning migrants that I couldn't imagine ever having too many birds—even when Maggie announced that she was moving nearly an hour away and that I would probably be getting all her bird calls.

The phoebes had returned, as always, on April 1. Though not a strikingly colored bird, the eastern phoebe is irresistibly jaunty, spending much of its time energetically searching for flying insects, bobbing its tail, and whistling its signature two-note "phee-bee! phee-bee!" Our resident phoebes' nest sits on a small platform under one of the eaves of John's office, safe from weather and the larger predatory birds. When I hear the first phoebe announce its cheerful return I give a gasp of delight: once again they've survived their perilous migration, and spring can't be far behind.

Many birds travel thousands of miles twice a year, and thanks to the rapaciousness of real estate developers, each year their journey becomes more hazardous and difficult. Still, they are amazingly consistent when it comes to arrival and departure dates, although no one is quite sure how they manage to do it. The rose-breasted grosbeaks arrive at our house on April 29, spend a couple of days helping themselves to the plentiful seed supply, then continue north to their summer homes. At the beginning of my second year of home bird rehab I received what would become a standard springtime phone call.

"Can you help me?" asked the distraught voice. "I have the most beautiful bird in my yard but someone must have shot him! He's all bloody on the chest! Please, can you come over and get him?"

"Is he black and white with a big thick beak?" I asked. "Not to worry—it's a rose-breasted grosbeak. They're supposed to look like that."

"Ohhhh," said the voice. "What a relief! I was sure he was dying, and I couldn't figure out why he kept acting like he didn't care."

By May 1, the hummingbird feeders have been up for a week, just in case one of the ruby-throats arrives early. All summer they circle the house at warp speed, splendidly iridescent and outrageously pugnacious, challenging each other to duels and vigorously defending their feeders even though there are five from which to choose. Although they weigh less than a nickel, their battles are epic: my friend Jan recalls seeing two male hummingbirds locked in combat fall through the air and roll around on the ground, neither willing to relinquish its hold on the other, both too furious to pay any attention to the huge human standing beside them.

My first Canada goose arrived soon thereafter. The bane of golfers and soccer players, Canada geese are undeniably messy; like most birds, they have no reason not to be. I had always considered Canada geese to be handsome birds with voices like wind chimes, a stirring sight when flying in their V-formations. But I didn't really appreciate them until Ponie arrived with one in a box.

Ponie Sheehan truly loves Canada geese. Anti-geese arguments cut no ice with her; she counters with poetic descriptions of their elegance, their resourcefulness, and their devotion to family. At work in a congested Westchester area near a large pond, Ponie would watch the resident geese with trepidation as they dodged the cars speeding in and out of the parking lot. One day she spotted a goose limping heavily and falling over. Grabbing a large cardboard box, she raced out to the rescue.

I lifted the goose out of the box and put her down in our garage. She was alert and bright-eyed, only slightly thin, and unperturbed by the two humans handling her. She was unable to bear any weight on her right leg, which was swollen and hot to the touch.

"I'll get her X-rayed, we'll splint it, and you'll probably have her back in a few weeks," I said, watching Ponie's face register concern, delight, and then such open gratitude that I felt I should reassess my usually negative view of the human race. "Why do they call you Ponie?" I asked.

"My siblings wanted a pony, not a baby sister," she said with a grin.

Carol Popolow X-rayed the goose, concluded that it was a spiral break trying to mend, then immobilized her leg with a splint and a snowshoe. One round of antibiotics, one round of worming medication, a couple of rechecks, and in three weeks her bone would be healed. Meanwhile all that remained was to feed her, keep her quiet, and clean up prodigious amounts of goose doo. When I appeared to do my chores she would watch me quietly, limping obediently to the other side of the shed so I could clean out her pen, as if living in a clinic and recuperating from a broken leg were something she did all the time.

Ponie called a few days later. "I'm sorry to bother you," she said in her soft voice, "but I was just wondering how the goose was doing. Is she better? Do you think I might be able to come by and see her?"

Had the goose been any other kind of bird I would have come up with an excuse, as recovering wild birds need as little contact with people as possible. But although some Canada geese can be quite wild and fearful, many are practically domesticated; as long as you didn't try to hold her down, this one belonged in the latter category. "Sure," I said. "Saturday morning is fine."

"Oh, thank you!" said Ponie. "Really, I don't know how to thank you."

"You should be a rehabber," I said.

"Someday," sighed Ponie.

During the next few days of beautiful spring weather we accumulated five nestlings, a black-capped chickadee and a mourning dove; and John reminded me that it was goshawk nesting season by staggering through the door clutching the back of his head.

"Jesus Christ!" he gasped. "One of those birds you're always talking about got me from behind!" As it turned out, he had been running down a trail, minding his own business, when he was nearly leveled by what he thought was a good-sized rock. "There I am, ready to pound the guy who did it," he said, looking aggrieved, "and I can't reach him *because he's perched on a tree limb.*"

As it turned out, the goshawks had finally abandoned their old nest and built another one. From a goshawk's point of view, the new location couldn't have been better: high in an oak tree in a heavily wooded area, right beside a wide old carriage trail that provided a little space around the nest. From a human's point of view, the location couldn't have been worse: although few people used the park, those who did generally took that particular trail. I sympathized and exclaimed over John's bloody head, told him firmly that he should not use that trail again until the nestlings had fledged, then raced off into the woods.

I rounded a slight curve in the trail, my eyes trained on the tops of the trees, and there it was: a good-size, well-constructed nest built firmly in a high tree crotch. Standing sentry on the rim, to my delight, was the female goshawk. *I wonder if she remembers me?* I thought hopefully. We stared at each other until she

broke the silence with her slow chant, which built in volume and intensity until it became a war cry that propelled her off her nest and straight toward me.

The carriage trail was like a runway, wide and unobstructed, and for a hundred feet or so I had a clear view of her amazingly swift approach: the quick beat of her powerful wings, the way her streamlined body sliced through the air, the bright glare of her dark red eyes that locked into mine as she hurtled toward me. She was the most beautiful, graceful, lethal creature I'd ever seen in my life, and I was so awestruck that I almost forgot to duck.

But duck I did, and when she banked up into a tree, I whirled away through the forest like Julie Andrews in *The Sound of Music*, my rapture triggered not by a sunny day in the Alps but by an angry meat eater who had just tried to behead me. Everyone has her little quirks, I thought giddily as I tangoed into the kitchen.

"When can we go?" cried the kids after I'd regaled them with my adventure. "When can we see her?"

"Uhhhhhh," I stuttered, once again cursing my so-called parenting skills. "I'm sorry—I really am—but it's awfully dangerous, and we all have to leave them alone so they can have their babies."

"What?" they howled. "You never let us do anything!"

"Really!" said John, shaking his head. "Not letting a psychotic bird of prey maim your children—what's the matter with you?"

"That's right!" they wailed. "What's the matter with you?"

"Nobody can go there anymore," I said firmly. "I'm going to put signs up on both ends of the trail telling people not to hike there."

"Except for me," said John, "because that's my running trail."

"No way!" I replied. "You can go back at the end of July."

"I'll wear a bicycle helmet," he countered.

"A bicycle helmet!" I snorted. "Try a suit of armor!"

The following day I collected half a dozen SENSITIVE WILDLIFE AREA—DO NOT ENTER signs from the local Audubon center and posted them on both ends of the trail. I painted two of my own that announced DANGER—NESTING HAWKS and put them up for good measure. While the thought of a large pair of raptors chasing terrified people out of the woods filled me with misanthropic glee, I was fearful of what could happen to the birds if they scalped someone unsympathetic. Occasionally my concern would boil over at the wrong moment, as it did when John lurched through the kitchen door, blood flowing down his face from a wound delivered straight through the air vent of his bicycle helmet.

"Oh, my God!" I gasped. "What did you do to that bird?"

John gazed at me through narrowed eyes. "The day I finally arrange for your competency hearing," he said, "you'd better hope I don't bring this up."

If the female was the sentry guarding base camp, the male was the scout patrolling the perimeter. Although John and I both avoided the nest trail, occasionally I would encounter the male some distance from the nest. Quiet and stealthy, he would appear out of nowhere and rocket past my head, the tip of his wing a few inches from my ear. I'd hear a *ffffsssSSSHHOOOP*! and by the time I reacted—by jumping two feet into the air and stumbling off the trail— he'd be regarding me silently from a tree limb ten or fifteen feet over my head. Regaining my shattered equilibrium I would stare back and he'd let out a series of soft whistles, turning my heart to jelly and making me vow to kill anyone who so much as raised his voice around him. When I continued my run he'd fly past me once more, this time at hip level, then veer off into the woods.

After one of these encounters I returned home to find a yellow Toyota in my driveway. A petite redheaded woman smiled at me from the driver's seat, then opened the door and climbed out. Tanya turned out to be a rehabber who lived a half hour to the north.

"Hi!" she said. "I'm sorry to just appear like this—I've just come back from a vet in Westchester. I keep hearing your name, and I thought I'd stop in and say hello. So you do birds! And you have a flight cage!"

I showed Tanya around. "Your flight cage is awesome," she said when we returned to her car. "Listen, if I end up with birds, can I bring them to you? I'd rather not do birds if I can avoid it. Here—see? I have this grackle. Somebody found him on the ground. I had him X-rayed and there's nothing broken, but he's kind of unsteady. The vet said he probably hit a window and just needs some time. Would you want to take him?"

"Sure," I said. "I like grackles."

"Oh, thank you, that would be terrific!" said Tanya. "Here, let me give you some grapes. I just went to the store and he really likes them."

We set the grackle up in one of the reptariums in the shed, putting in two different-size perches high enough to get his long tail off the ground. He was an otherwise healthy adult, his glossy feathers appearing black at first but then flashing bronze, blue, green, and purple when they caught the light. By now Null and Void would look just like him, I thought.

"I'm so glad you're here," said Tanya as she was leaving. "People call me all the time with birds, and I just can't say no."

I had no idea how much those words would come to haunt me.

Chapter 30

TORRENT

I received an e-mail from Ed. "I was reading *The Merchant of Venice* again," he wrote, "and I came across this quote, spoken by Portia to Nerissa:

The crow doth sing as sweetly as the lark
When neither is attended, and I think
The nightingale if she should sing by day,
When every goose is cackling, would be thought
No better a musician than the wren.
How many things by season seasoned are
To their right praise and true perfection!

"This morning I was birding on Plum Island," he continued, referring to a wildlife sanctuary off the Massachusetts coast, "and with every bird I saw I thought: 'by season seasoned are / To their right praise and true perfection.' I expect that phrase will soon be lodged in your head, too, if it isn't already. By the way, have you any of the above?"

"Thank you for the poem," I wrote back. "And yes, as a matter of fact, I have a goose, although she doesn't do much cackling. But none of the others."

By one of those small but strange coincidences, less than twenty-four hours later my statement was no longer true. Tipped off by Tanya, a rehabber named Kelly called and asked if I could take two nestling crows.

Everyone wants a pet crow, but luckily for the crows, it is illegal to keep one without a special license. Crows, like parrots, are very intelligent, intensely focused flock birds who hate to be alone. A single crow (or parrot) raised from a nestling will want to be with its foster human constantly and will suffer greatly if ignored. A crow cannot be raised by people and then, when it gets to be too messy and time-consuming, suddenly set free; it will have no idea how to relate to other crows, find food, or recognize predators, especially the human ones. Young crows have to be raised with other crows. Luckily for me, these two came as a pair.

"Sure, I can take them," I said, after she told me there was no way to get them back to their parents. "Can you bring them over here?"

I didn't add, "within the next five minutes?" although I wanted to. It had been years since I raised my pair of crows, and the one who survived had long since joined the wild crow flock and disappeared. The one who died remained, buried on a grassy slope behind our house, never having relinquished her fierce hold on my heart. Because of her I watch wild crows, follow them through the fields when I go running, and gravitate to the unreleasable ones at nature centers, always marveling at their intelligence and their complicated personalities. Crows are too smart for their own good and have no sense of law and order, traits that make them annoying to many people but irresistible to me. They're bullies and opportunists, travel in gangs, and harass better behaved birds, but they're also loyal and affectionate and like to slide down snowbanks on their backs. When the two nestlings arrived, they gazed up at me through wondrous blue eyes, and I felt a small poignant stab of remembrance.

They were thin and their feathers a bit ratty, but otherwise they were healthy young crows. They were pre-fledglings, covered with fuzzy dark feathers and not quite ready to leave their nest. While much of crow behavior is instinctual, even more is learned from parents and extended family. Nestling crows are trusting, as their family has not yet clued them in on the human penchant for assault, murder, and mayhem. But adolescent and adult crows are well aware of it and can be difficult birds to rehab, simply because they are so terrified of their captors.

Since their hourly feeding schedule precluded them from going into the shed, I brought a medium-size carrier into the spare bathroom, lined it with newspaper, and outfitted it with a cozy nest. I added a small log, just in case they felt inclined to hop out and move around. Soon after they arrived I approached them holding an appetizing bowl of crow food—soaked puppy chow, hard-boiled egg, raisins, peanuts, vegetables, fruit bits, mealworms, chopped mouse, and vitamins. Seeing my extra-large pair of tweezers they stood up, opened their beaks, and emitted a series of loud begging cries, eagerly gobbling down each offered bite.

I felt a surge of emotion. Stop it, I told myself sternly. You are a professional. There will be no more falling for any crows around here.

The number of bird-related phone calls increased, and people called from farther away. Occasionally I would open my door to find a stranger holding a wounded fawn, or receive a telephone plea to help an injured coyote. The shed slowly filled with injured birds, the bathroom with nestlings. If I had managed to juggle four balls the previous summer, this summer I was juggling eight balls, three torches, and a rake.

"Dear Marigoldy," wrote Skye. "What will you be doing for Summer Solstice? Will you and your fairy friends have a party? Can you tell me who will be there, and what kind of food you will eat?"

Skye's fairy notes were always on my daily To Do list, although they were always the last to actually be accomplished. Rising at 6 A.M., I would head for the kitchen, make myself a cup of tea, then head off to the spare bathroom for the first nestling feeding of the day. I would check the critical-care birds, if I had any, steeling myself in case they hadn't made it through the night. Returning to the kitchen, I would soak puppy and kitten chow and chop fruits and vegetables for the birds in the shed and the flight cage, finishing up with an array of dishes filled with various enticing food combinations. I'd get the kids up, feed them breakfast, and get them dressed and ready in time for John to come in from his writing cabin and take them down to the school bus. I would spend the day feeding and caring for birds, fielding phone calls, giving people directions to my house, and accepting the wounded and orphaned, somehow sandwiching house and family chores and errands in between.

When the kids jumped off the bus at 3:20, I'd take them to the pool or to the store, usually lugging the nestlings along with me. Then there was homework or, during the last few weeks of school, an unending series of activities. Odds are Tanya would have called sometime during the day, and would arrive with a wounded bird who needed immediate attention right around 6:00. This was the time she finished work and was driving past my house, but it also happened to be when the family dynamic had reached its crescendo: the kids were hungry, the parrots were screaming, and John was walking in from his cabin with the pipe dream of enjoying a leisurely beer on the deck.

"I'm sorry," I'd say to him. "I'll be as quick as I can."

"Yeah, yeah," he'd sigh, and start to prepare dinner.

Occasionally we all went outside after we'd eaten, or watched a movie, but inevitably I would be called away by a phone call or the arrival of an injured bird. When I put the kids to bed, John and I could finally spend some time together, then about 11:00 or so I'd suddenly remember the Marigoldy notes and have to race off to find paper and colored markers. I'd slip the finished note under Skye's pillow, yawning and wondering how long this phase would last; but when she read the note the next morning, giddy with delight and awed by the magic in the world, I hoped it would last forever.

I had never worked so hard in my life, and the weekends were worse than the weekdays. But like Skye, I too was awed by the magic in the world, the kind that allowed me to watch a bird grow from a tiny naked nestling to a healthy fledgling, or to see an adult bird recover from an injury that would otherwise have doomed her. At times my learning curve was so steep it felt almost vertical, and I often despaired over my own ignorance. Once I posted yet another question on one of my electronic mailing lists and followed it with an emotional declaration of incompetence, triggering a flood of e-mails filled with support, generosity, and kindness.

"Don't start getting crazy on us," wrote one veteran rehabber. "I've been doing this for thirty years and I still don't know what the heck I'm doing."

"You're doing a great job," wrote another. "Hang in there and try not to stress out too much."

"I don't know what I'd do without all of you," I wrote back.

"Don't tell us you love us," cracked another. "Just send money."

There were complications with many of the summer's nestlings. There were some whose parents, knowing they had such serious defects that they could never make it on their own, probably pushed them from the nest. There was one small nestling whose head was twisted to the side; when it gaped for food, its head turned upside down. Another's back was hunched, its legs crossed and feet curled. There was one with missing toes and half a wing, another whose head moved slowly back and forth, like a metronome. Had I received two from the same nest, or even the same town, with similar problems I would have reported it to the Department of Environmental Conservation, as perhaps there might have been a provable connection to heavy pesticide use in that area. But they came in one by one, over a period of two years, and all from different directions.

There were the otherwise healthy ones who were victims of human interference. A sparrow whose leg, tightly encircled by a piece of string the parent had used in making the nest, came off when I tried to free it; the fuzzy little almost-fledgling whose wing had been ripped off by a cat.

It fell to me to let them go. When the healthy young nestlings were released, they would live out their lives as wild birds. If any of the others survived, they would spend their crippled lives in a cage. There was not a ghost of a chance that they could someday live the way they were meant to live.

I couldn't let a suffering bird languish in a box until he died, as it could take hours, even days. The first time I dug a small hole in the woods and gently laid a mortally wounded nestling down, his obvious pain sent my adrenaline surging and allowed me to do what I would never have thought I could. Placing the heavy shovel above his neck I whispered, "I'm so sorry" and brought it down, stepping on the edge to make sure it was over. I filled in the hole and then sat beside it, trembling, my face in my hands.

It never became any easier.

Chapter 31

THAT STRANGE CHIRPING SOUND

My father arrived on the 5:34 train from New York. After leaving his home in Colorado, where he had moved after my mother died, he spent a few days in the city with old friends and would be going on to visit my brother after three days at our house. When Mac and Skye were toddlers they had somehow transformed "Grandpa" to "Grumpy"; this inspired us to dub Missy, the tall, sardonic blonde with whom he had lived for ten years, "G.G.," or Grumpy's Girlfriend. Dad would entertain the kids with arcane scientific facts and math puzzles, G.G. would read them children's books in French, and we'd all go for a hike in the woods. But G.G. was traveling elsewhere this time, so Dad arrived on his own.

He greeted us heartily, disappeared upstairs with the kids for half an hour, then wandered through the living room. "Everything looks very nice," he murmured appreciatively, even though nothing had changed since his last visit; actually, not much had changed since we moved in ten years before. "And look!" he continued, as if he were admiring a particularly intricate new piece of stereo equipment. "You even have maggots in your dining room!"

I grew up in a world with a calm and gracious exterior. Everyone my parents knew had a sense of decorum; at least, they did until the early 1970s, when social mores started to unravel and things got slightly out of hand. But even then, kids went to dancing school, women wore evening dresses to each other's parties, and if a man talked openly about the cost of his possessions he immediately became a social pariah. Conversation was an art, and overreacting—to *anything*—was a hanging offense. Had my mother walked into the dining room and found a pair of vultures standing on the table, she would have smiled and said, "It's a good thing your grandmother isn't here—you know she wasn't all that fond of birds."

Dad had left much of his old life behind, but held on to certain inviolable traditions: cocktail hour, for one. Soon we were on the deck, Dad drinking vodka on the rocks while John and I, who had many times experienced the painful consequences of trying to keep up, consoled ourselves with glasses of wine. The kids busied themselves with cheese and crackers and pointed to various birds traversing the slope—three robins, several different types of sparrows, and a pair of jays—who might have been birds we released last summer. Dad expressed interest in all our updates, alarm at my nestling-feeding schedule, and after inquiring as to what sort of bird paraphernalia I was lacking, offered to build me an outdoor cage.

"No problem," he said. "Just a nice big cube with a door. Something you can pick up and move around, but that will protect the birds inside. No bottom, right? Why clean it if you don't have to?"

The next morning, as my father sat drinking coffee and making what looked like a series of professional architectural drawings, I tried to figure out how I could possibly spend time with him as well as accomplish all I had to do. There were birds to care for and nestlings to feed; during one of my thirty-minute breaks I had to rush into town; buy a birthday present for one of Mac's friends; stop at the drugstore; pick up the live crickets, mealworms, and waxworms waiting at the post office; then race back in time to drive Mac to the party. When I returned, I decided, I could finally spend some time with my father, maybe even help him build the cage. I was negotiating with Mac to feed the nestlings if I was delayed when the phone rang.

"Suzie?" said the voice. "This is John Lucid down at the post office. I'm afraid we have a situation here."

"A good situation?" I said idiotically.

"Not according to Karen," said John. "Your box of crickets arrived damaged, and, uh, it seems a few of them got out. Karen's afraid of bugs, and they've got her cornered up against the wall."

"No!" I said, my mind racing: how could I possibly fit a cricket roundup into my schedule? "I'll be right down," I said, and hung up.

I hurried into the post office, wondering how many of the 1,000 crickets I'd ordered had escaped. Twenty? Thirty? Karen was standing behind the counter, wide-eyed; from the room behind her came a melodious chorus of chirps.

"Oh, my God," she said. "They're everywhere! I told John I'm either staying up here in front or I'm going home. It's like a horror movie back there! I'm going to have nightmares for months!"

"I'm so sorry!" I told her. "I'll get them out of here as fast as I can!" I disappeared into the back, sure Karen was exaggerating.

She wasn't. The back of the post office was dark with crickets: strolling along the floor, hopping out from behind large packages, surveying the room from the tops of mailbags, all waving their little antennas in a friendly manner.

"Pretty funny!" said John, chuckling to himself. "Good luck catching 'em all!"

John had placed the damaged box containing the crickets that hadn't escaped into a large mailbag. I borrowed another one and started chasing the rest down, soon realizing that compared to a songbird—even a juvenile songbird— I was astoundingly clumsy. Could I return home, I wondered, gather up my birds, return, and let them catch all the crickets? My deliberations were interrupted by an anguished gasp from the counter in the front.

"She's catching them *with her hands?*" Karen hissed. "Oh, my God. The *nightmares.*"

En route to driving Mac to his birthday party, I explained why I had been delayed. "You got to catch all those crickets yourself?" he cried. "Why didn't you come home and get me? How come you get to have all the fun?"

When I returned Dad had taken his list and driven himself off to the hardware store in John's car, leaving me to feed the nestlings, run Skye to a friend's house, and do a few assorted chores. He returned with several bags of supplies, set himself up in the garage with a radio and a tall glass of iced tea, and created what we would eventually christen the Crow Mahal. Portable and made of a lightweight skeleton surrounded by hardware cloth, it would allow young crows to enjoy the grass and the sunshine in a secure and hawkproof environment. Whenever ducklings arrived, I could remove the perches and presto! The Duck Mahal. When Dad was nearly done I appeared, breathless, finally able to spend some time with him.

"This is quite an operation you're running," he said. "Are you sure you're okay?"

"Oh, absolutely," I said, as a yellow Toyota pulled up behind us. Tanya climbed out, and I introduced her to my father.

"Nice to meet you!" said Tanya. "I'm sorry—I tried to call you and there was no answer. I was passing right by and figured I'd just see if you were home. Would you be able to take an almost-fledgling starling and a sparrow who was caught in a glue trap?"

This isn't fair, I thought with anger and frustration. *I see my father twice a year.* "The starling is no problem," I said. "I have one about the same age and I'd be glad to put them together. But I'm kind of swamped—can anyone else take the sparrow?"

Tanya shrugged helplessly. "Go ahead," said Dad. "I'll finish up here, and meet you inside."

"All right," I said. "Poor sparrow."

There must be a special place in hell for the person who invented the glue trap—no doubt right next to those who invented poison, the harpoon gun, and the leghold trap. Glue traps don't deliver a quick death; they simply hold the animal fast until it starves to death or rips one of its own limbs off trying to escape. The glue is a particularly disgusting mixture; one must dissolve it with surgical tape remover, then bathe the bird in Dawn dishwashing liquid, the savior of oil-spill-contaminated wildlife. And although wild birds love to bathe, they don't appreciate *being* bathed, so removing anything from a bird's feathers is a complicated process, terribly stressful for the bird, and usually must be done in a series of attempts.

The little female house sparrow had ripped out several feathers trying to escape from the trap. I removed the glue, bathed her, made sure she was warm and dry, and put her in a carrier in my bathroom. Before leaving I opened the door once more to check on her and she came barreling out, blew past me, and disappeared under the bed. Cursing myself for not closing the bathroom door before I opened the carrier, I felt around under the radiators, pulled everything out of my closet, and rechecked the bathroom. Nothing. I couldn't find her.

I tried to get ahold of myself. It was cocktail hour, for Pete's sake. I hadn't been able to take my dad for a drive, or out to lunch, or even accompany him to the hardware store because of all these birds, and he was leaving in the morning. I could hear him laughing with John in the kitchen, and the kids running around outside. With any luck the sparrow would return for food and water; I brought the carrier into the bedroom, propped its door open, and after carefully closing the bedroom door behind me, headed for the kitchen.

After dinner we were playing charades in the living room when Dad looked up. "Do you hear a cricket in here?" he asked.

"I swear they're all out in the garage!" I cried. "I put rocks on the aquarium tops so they couldn't escape!"

"Mac and Skye each have two of them in their rooms," said John. "The last time we had a cricket delivery, they convinced *their mother* to let them have a pair as pets."

We looked at the kids. Mac appeared studiously neutral, while Skye wore the exaggeratedly innocent look of a child actor from the silent film era.

"What did you do?" I asked her.

"It wasn't my fault," she said. "I was teaching Cyrus how to balance on my hand and he jumped off. And then I tried to catch him and he ran away."

"So he's loose in the house somewhere?"

"He's not the only one," said Mac.

Skye glared at him. "I didn't want him to be lonely so I let Esmerelda go, too."

"Yeah!" said Mac. "Then she came into my room and tried to let mine go!"

"What?" said John. "Why would you do that?"

"Because it wasn't fair that two of them got to go and two of them had to stay!" said Skye. "If one was already gone, what's the difference between one cricket and four crickets? That's what Mommy's always saying! 'What's the difference between one bird and four birds?'"

"But Mom's birds aren't loose in the house!" retorted Mac. "Right, Mom?"

I looked from face to face, not about to confess solidarity with my slippery-fingered daughter.

"I wouldn't worry about it," said Dad. "They won't eat much."

After everyone had gone to bed I crept into the bedroom. The sparrow wasn't in the carrier, but she had been there earlier, judging from the seeds scattered around inside.

"I'm sure I'll find her tomorrow," I said, after trying to explain the situation to John.

The following morning I lay curled on my side as the first rays of sunlight crept through the blinds. Drowsily, I opened my eyes. Perched comfortably on my leg was the sparrow, preening her beleaguered feathers. Gathering myself for the assault, I sat up suddenly and threw the lightweight blanket over the bird, who raced out from under it and sped off toward the bathroom. Throwing myself out of bed, I thundered after her, leaving John shouting, "What? What happened?" behind me. Eventually I cornered her and a moment later emerged triumphant, sparrow in hand.

"Got her!" I said, grinning.

"Great," said John, covering his head with a pillow. "On to the crickets."

After breakfast I carried a sheet-covered reptarium to the slope outside the kitchen window. A little chickadee was ready to go, there were a number of others in the vicinity, and I wanted to release a bird in my dad's honor. As

John and the kids watched from the deck, Dad unzipped the reptarium and the chickadee flew off to the safety of a juniper bush.

"We'll watch out for him," called Mac. "And we'll name him Grumpy."

As Dad went upstairs to pack his clothes, I cleaned the reptarium with a scrub brush, disinfectant, and the hose and left it to dry in the sun. I had just finished feeding the nestlings when I heard Skye's voice from the kitchen. "Hey!" she shouted. "Grumpy's in the reptarium!"

On my way to the kitchen I had several moments of unlikely visuals, all involving my father somehow folding himself into a small mesh cage. I joined Skye at the bay window, however, and found the little chickadee hopping jauntily about the scene of his former—I thought—involuntary confinement. My dad came in and looked over my shoulder.

"You run a nice hotel," he said. "None of us want to leave."

Chapter 32

THE LUCKY ONES

"It's a mess," said Alan. "Old enough to fledge but it's neurologic. Spinal damage. Emaciated. There's nothing I can do. I advised her to euthanize it, but she didn't want to hear that. She's on her way to you now."

"Oh, damn it, Alan," I said. "I have so many birds, and most of them come from her. She's driving me nuts."

"No doubt," said Alan.

When Tanya climbed out of her car, I was ready for her. "Tanya," I said. "I cannot keep handling all the birds you bring me. There aren't enough hours in the day."

"What am I supposed to do?" she replied.

"You live thirty minutes north of here," I said. "There have to be other rehabbers closer to you! You've got me doing three counties' worth of birds!"

"There aren't enough bird rehabbers anywhere," she said. "And when people call me, I can't just tell them no."

"Then you're going to have to start taking some yourself," I said, exasperated.

"Suzie," she said in a why-must-I-state-the-obvious tone, "I have a full-time job, and you don't even work."

I stopped, so thrown by her statement that I couldn't think of a response. "I 'don't even work'?" I repeated.

"Of course, you do," she said hurriedly, probably alarmed by my expression. "I mean you don't work nine to five."

"Even with a full-time job," I said, "you have way more time than I do!"

"Fine," she said. "It's all right. I'll take the crow to work with me and leave him in the car. Or maybe my boyfriend can help me."

"Alan called me," I said. "He said the crow is a mess. Why didn't you let him euthanize him?"

"Because he deserves a chance," she said, and turned to get into her car.

Let her go, I thought. "Let me see him," I said.

He was painfully thin, his feathers dry and tattered, feet clenched, legs curled and unmoving. His head nodded unsteadily, and his tongue lolled from his beak. His blue eyes were dull.

I set him up in my bathroom, away from the other birds. His nest was a heating pad in a wicker basket covered with a terry cloth towel, then topped with a layer of paper towel for easier cleaning. I tubed him with a tiny amount of a mixture designed for patients with impaired gastrointestinal function, then left him, hoping he would sleep.

I closed the door behind me. I could continue to feed him tiny amounts of nutrients and possibly put some weight on his wasted body, but I couldn't fix his brain or his spine. If he survived the next few days I could spend weeks feeding and providing him with physical therapy, but the odds were over-whelming that I would end up with a crow who couldn't walk, fly, or feed him-self. He wasn't a handicapped human cared for by other humans; he was a flock bird with no flock, his brain damaged not by possibly repairable trauma but by genetics gone awry.

A skilled veterinarian had told me there was nothing he could do; there was nothing anyone could do. I tried to force myself to accept the obvious, to simply be grateful that I could make the little crow as comfortable as possible during the time he had left, but I was buffeted by waves of helpless frustration. Why? I wanted to shout. *Why can't I help him?*

My logical self recognized my limits. But a small, hidden part of me took each hopeless case as a personal failure.

Impatiently pushed aside, then carefully tucked away.

Out in the yard our two young crows were enjoying the sunshine in the Crow Mahal, where they spent a few hours per day. They were healthy and active, had abandoned their nest in favor of perches, and were almost eating on their own. They were endlessly curious. Delighted by their ability to be awed by a pinecone, we had christened them Lo and Behold. Protected by Dad's square bottomless cage, they picked through the grass, splashed around in their water dish, and invented games to go with whatever I tossed inside.

Keeping a crow in a bare cage is the psychological equivalent of keeping a prisoner in an empty cell. Crows need sticks, leaves, small logs, molted feathers, pebbles, berries, flowers, weeds, roots, and seed pods, just for starters. Each one will be analyzed, fought over, thrown about, torn apart, smeared with food, and finally dropped into the water bowl. There are few creatures messier than crows, but then, there are few creatures who *enjoy* being messy more than crows.

I wished I could sit on the grass, lean against a tree, and just watch them play, but I had too much to do. I did another round of chores, then returned to the new crow. After I gave him a second small amount of tubing mixture I started to put him back in his nest, but stopped and instead held him quietly on my lap. Had he become agitated I would have put him back, but he settled down, leaning against me. He raised his head, beak agape, tongue flicking in and out. He glanced up and then his head dropped, shuddering from side to side.

I brushed the feathers on the side of his cheek with my fingers and followed them slowly to his throat, so lightly it was barely a touch. Continuing across to the other side, up to the ridge of bone above his eye, along the top of his skull, and down the nape of his neck. The tremors subsided until they were ripples on a quiet pond. His eyes closed.

Crows live in extended families. Young crows stay with their parents for three to five years and help feed their younger siblings. Even if a young crow moves away from its nuclear family it may return for occasional visits, and often it moves no farther than a few miles away. If a family member is on the ground, injured and unable to fly, the others will feed and care for it. Abandoning a fledgling is something a crow family would never do.

Unless the fledgling had brain damage and partial paralysis.

I cupped my hand around the little crow, hoping to provide some kind of comfort. Tomorrow I would bring him to Wendy, to do what his family had no choice but to do. Until then, I would keep him warm and fed and comfortable. Blinking back tears, I settled him in his nest. I didn't have time for grief; there were other birds who needed care.

Before I went to bed that night I quietly opened the bathroom door and found him sleeping, his small form cushioned and still. I dreamed of a flock of crows so enormous it approached our house like a summer storm, blackening the sky, all of them protectively encircling one small individual.

"You're flying," I shouted to the crippled crow, as bright bolts of iridescence shot across the underside of his wings like heat lightning. "I can see you flying!"

In the morning I found him in the same position in which I had left him. "I'm sorry," said John. "You were going to put him to sleep anyway, though, weren't you?"

"Yes," I said, my voice breaking.

"I'll take the kids to camp," he said. "You do whatever you have to do."

I cleaned, I fed, I medicated. I decided Ponie's goose needed a bath and dragged a large plastic kiddie pool into the parrot's flight, filled it with the hose, and carried the goose in. She eyed the pool suspiciously for a few minutes, then poked her beak into the enticing water. She hopped in, rustled her feathers, and kicked her feet; she ducked her head under the water several times in a row and let the rivulets run down her neck. She spread and flapped her wings, then thrust her head under the water and dabbled. She shook herself, opened her beak, and let out a long hiss, like a dame from the 1940s exhaling a luxuriant trail of smoke.

My goose reverie was interrupted by the ring of the telephone. "Hi," said the voice. "It's Erin Smithies from Teatown. I hear you have crows. Would you be able to take a fledgling?"

"What's the matter with him?" I asked, my voice more harsh than I meant it to be.

"Nothing," she said. "He's fine. Just a little thin. I came to work a few days ago and he was in a box on our doorstep. No note or anything. I feel so badly for him—I'm sure he has a family somewhere."

Teatown Lake Reservation is an 834-acre nature preserve and education center. Their executive director is Wendy's husband, Fred Koontz, a biologist and biodiversity expert who has traveled the world while studying and working with wild animal populations. As a rule, local animal people are all connected. Everyone knows everything about everyone else. I know Fred and I know Erin, both of whom are professional, dedicated, and conscientious; one might think that if Erin said there was nothing wrong with the crow, I would believe that there was nothing wrong with the crow.

But as the day went by I became more and more gloomy, and by the time Erin pulled into the driveway I was sure the crow in her carrier was blind, deformed, and riddled with pox. I managed to return Erin's smile, though I kept glancing suspiciously at the carrier.

"Can I take a look at him before I put him outside?" I asked. "We can go in the garage."

I put the garage door down and the light on, and Erin carefully removed the crow from the carrier.

He was beautiful.

"He's awfully easygoing," she said. "Didn't take long to settle in—which is why I wanted to get him with other crows as fast as possible."

I led her to the backyard, where Lo and Behold perched in the Crow Mahal. It could easily accommodate three young crows, so I placed the new crow inside. All three assumed various positions, the two siblings on one side and the new crow on the other, all looking quite taken aback.

"Will you let me know how he does?" asked Erin. "He's such a nice crow."

"I will," I promised. "I'll release them here, so they can always come back for food. They'll be fine together."

Later that afternoon we all gathered on the deck to watch them before I brought them inside for the night. They were like kids in a schoolyard, the bond between siblings unmistakable but easily stretched to include a playmate. We named the new crow Nacho, in honor of the fact that John, the kids, and I were actually going to a Mexican restaurant for dinner that night.

At that point in time my leaving the house for a night out was akin to climbing Mount Kilimanjaro. If I was overloaded with nestlings, it was like climbing K2. An overload of nestlings plus several critical-care adults equaled Mount Everest. If Tanya had just pulled in, it was like flying to the moon with no rocket.

But on this particular night I was determined to get out, and to get out on time. I had been feeling more and more guilty about my constant nestling-feeding schedule, as well as my inability to do anything fun and spontaneous with my family. I was bustling around, twenty minutes away from leaving the house, when the phone rang.

"Don't answer it!" I shouted, but not in time.

"It's Karen Parks," said Skye. "You won't be long, will you?"

I took the phone with a sigh of relief. Karen lived a few miles away; John played pond hockey with her husband, Steve. If I know the person, I somehow reasoned, they can't have an injured bird for me.

"I'm sorry to bother you with this," said Karen, "but a big red-tailed hawk got into our chicken run and killed one of our chickens, and I think she broke her wing. The females are the big ones, right? She's on the ground and she can't fly. Steve went out with a blanket and put her in a big cardboard box. Can I bring her to you?"

"Uhhhhh," I said, "I, uh, um. Yeah—it's okay, bring her over. Can you come right away?"

I hung up the phone and found all three members of my family staring balefully at me.

"You're not . . ." John began.

"No, I'm not," I said loudly. "Karen's going to drop off a redtail, I'll wrap her wing, put her in the crate in the garage, and she'll be fine until morning."

"Promise?" said Mac.

"Promise," I said.

Ten minutes later I carried the cardboard box into the garage, pulled on a pair of gloves, and opened the top. A very large redtail hissed, flipped over on her back, launched herself upward, and tried to grab me with two formidable sets of talons. I snatched at her legs and caught one; in the ensuing struggle it became apparent that there was nothing wrong with either her wings or her legs. I pinned her wings against my arm, she seized one of my gloves in a death grip; locked together we staggered over to the crate, where the question of who would let go first waited to be answered. Realizing it wasn't going to be her, I pushed her into the crate at the same time I pulled my hand out of the glove. For a moment she stood, angrily squeezing the empty glove. Finally she let it go and hopped up onto the perch, shaking her feathers in outrage.

The shadows created by the single ceiling light made it difficult to see into the crate. I could see the hawk's silhouette, though, standing square with wings tight against her body, and I could certainly feel the waves of fury she sent crashing into my face.

"Keep the glove," I told her. "I'll see you in the morning."

We ate guacamole and salsa, tacos and enchiladas. John and I drank margaritas and Mac and Skye ordered huge banana-and-ice-cream concoctions for dessert. The kids were giddy, John was jovial, and I felt as if I were playing hooky, barely able to recall when I had taken evenings like this for granted. I remembered when I announced that I would combine bird rehab with raising young kids, and my rehabber friends had dissolved into laughter. Perhaps I had been a bit hasty in dismissing them.

The following morning I donned my gloves and reached into the redtail's carrier, determined to find the mysterious malady that had grounded what seemed to be a healthy hawk. I pulled her out, her eyes filled with rage and one set of her talons embedded in my glove, and stopped. How could I have missed it?

"Karen?" I said into the telephone.

"I'm so glad you called," she said. "We've all been so worried about the hawk. Is she all right? Is her wing broken?"

"She's fine," I said. "You know what a bird's crop is? It's where they store food until they can digest it. It's at the base of their throat. Hers is enormous. She couldn't fly because she's a big hog and ate too much chicken."

Wild hawks can't order room service. Their dinner is usually racing away from them as fast as it can, and a split second can mean the difference between a fine meal and no meal at all. It's easy to see why a raptor might overindulge, especially if it had been a lean few days.

"It's the Flyaway version of *La Grande Bouffe*!" I said to John. "Remember that movie? With Marcello Mastroianni?"

"The one where four guys decide to eat themselves to death?" he said. "Right! Back then it was social satire. Now they've remade it as *Super Size Me* and it's a documentary."

I kept the hawk until late afternoon to give Karen time to raptor-proof her chicken pen. Later I drove the recently christened Marcella to a sunny hill overlooking a field dotted with Black Angus, a mile or so from the Parks's house. As I opened the carrier door, the big redtail glared at me.

"*Essere attento, bella,*" I said. "Stay away from chicken pens."

I moved away from the door, and she bolted out and up into a huge old sycamore tree. She took in her surroundings and rattled her feathers, as if shaking off the shackles of her brief captivity. Easing off the branch she flew lazily due east, straight back toward the land of the easy pickings.

Chapter 33

SONGS OF GRIEF

"You'd better sit down," I said to Ed, "because I'm about to let somebody have it, and you had the bad luck to answer your phone."

"Oh, dear," said Ed. "Sounds like trouble in paradise."

I started right in. There were the last three callers, all of whom had found baby birds and decided to let their kids raise them—at least, until it was obvious that the nestlings were at death's door, which was when they brought them to me. There was the woman who let her daughter keep an injured pigeon in a bare box without food or water for four days because she had been "far too busy" to find it help. There was the man who called from his house an hour and a half away and demanded that I drive over immediately and retrieve the injured hawk in his yard, otherwise he would "get out his gun and shoot the damned thing." I ranted on and on about the stupidity, the callousness, and the treachery of the entire human race, myself included, until I ran out of breath. There was a moment of silence.

"Well," said Ed solemnly, "that will teach me to answer my telephone."

Once again, Ed talked me down. "First things first," he said. "You must keep in mind the species with which you are dealing. As Anthony Burgess said, 'Human beings . . . unsatisfactory hybrids, not good enough for gods and not good enough for animals.' No need to elaborate.

"However, do keep in mind all of the people who dropped what they were doing and went far out of their way to find help for an injured bird. All the people who managed to locate your house, which I can attest is in the middle of nowhere, and who called repeatedly to find out how the bird was faring. All the people who have donated money to your cause. And—most important— remember that one particularly noble soul who manages, time and again, to calm you down when you call him cursing and shouting, or write him entire e-mails using nothing but capital letters."

After Ed had whittled my stress down to a more manageable level I hung up the phone and sat in the kitchen, staring blankly into space. John walked through the door, stopped, and frowned at the sight of the large female mal- lard sitting under my chair. The mallard stood up, gave him an appraising glance, then sauntered past him and into the living room. After giving her tail a friendly wag she settled down between the kids, who were watching TV.

"Damn!" I said, slapping my hand against my forehead. "Here I have this big therapy session with Ed and I forget to bring up Danielle!"

"Danielle?" said John. "Is *Danielle* housebroken, by any chance?"

"No," called Mac indignantly. "That's why the lady kicked her out."

"And that's why," finished Skye with a sunny smile, "she's going to live with us."

Danielle had been found as a tiny duckling by a woman who brought her home, lavished attention and affection on her, and kept her in the house. When the duck was almost full size, the woman decided that since her housebreaking methods were not working, it was time for the duck to go. Of course, by this time the supposedly wild duck was completely imprinted and had never seen another duck.

I had made a half-hearted attempt to put Danielle in the shed, but I felt as if I were throwing a formerly pampered pet into a pound. "Look at her," I said mournfully to John, as Danielle waddled up and dabbled his pant leg with her beak. "It's not her fault she thinks she's a poodle! Okay, she might be a little messy, but just give me a couple of days until I can find her a good home."

"A little messy!" John exclaimed. "A full-grown duck? What are you going to have living in here next—a swan?"

"But Danielle is a special case, and she's very short-term!" I said. "She can spend most of the day outside, and I'll put her in a crate at night. She just needs a few hours a day of human interaction!"

"She just needs human interaction!" chorused the kids.

"Right," said John. "We certainly can't have the duck going cold turkey."

Ponie arrived on Saturday morning, carrying a tin of homemade cookies and wearing a huge grin. "I'm so excited!" she said giddily when John opened the door. "The goose is all better and going home! Isn't that the coolest thing?"

"It's very cool," John agreed. "Come on in. Don't trip over the duck."

Ponie followed me out to the shed, where she gathered up her goose and carried her to a waiting box in the back of her car. "You saved her life," she said. "I love what you do. Do you need a volunteer? If you need help I could come after work during the week, or for a few hours on weekends. I'll help you any way I can."

After she left, I wandered out to the flight. People always asked why I didn't have volunteers to help me. I certainly needed a break: when the kids were safely at camp there were times I simply burst into tears from sheer exhaustion. But coordinating volunteers would mean adding another layer of work, another level of responsibility. Ponie would be the volunteer from heaven: compassionate, inspired, easy to be around. But she could come only at 6 P.M. on weekdays—when most of the bird work was done and I was trying to concentrate on my family—or on weekends, which were a hectic blend of bird care and kids' activities. What if Ponie was scheduled to arrive just when the last nestling had been fed, I had already cleaned the crates, and I had a perfect half hour to play a game with the kids?

I was too tired. I couldn't figure out how to do it.

I reached the first flight cage. This particular group of songbirds consisted of three robins, a Carolina wren, two mourning doves, a northern waterthrush, the formerly glued house sparrow, and a white-breasted nuthatch. When I walked through the door the songbirds either froze or hid; all but the little nuthatch, who gazed at me from a dead limb, hopped onto one of the mesh-covered sides of the flight cage, then without hesitation kept going, creeping along the ceiling upside down until he was a few feet away.

"*Henk, henk,*" he said.

"*Henk, henk,*" I replied.

I opened the side door and entered the crow flight, closing the door carefully behind me. I had taken many of the smaller branches out, leaving the crows with more space in which to fly and solid branches on which to land. The enclosure was filled with all kinds of natural toys and a huge outdoor plant saucer for water. I sat down on a log and Nacho coasted down from a perch and landed beside me. When I picked up an acorn and tossed it into the water dish, he hopped after it, plucked it out, tossed it aside, then crouched down and began splashing energetically. Perhaps sensing a potential water fight, Lo and Behold left their perches and crowded around the dish.

Although they were still comfortable around me, Lo and Behold had stopped actively seeking physical contact and interaction, as is the norm when young crows are put into a flight cage before their release. Nacho, however, continued to be as easygoing as Erin had first described him. He had a solid bond with the other two crows, but was happy to follow me around the flight cage, offer me toys, and yank at my shoelaces; in turn I was happy to bring him extra crickets, toss him grass stalks, and give him the occasional head massage. I kept pulling myself back from the edge of infatuation, glad that I had only a limited amount of time to spend with him.

As I rose to leave I noticed that one of the knots holding a perch, a tree limb suspended from the ceiling by two heavy ropes, was becoming loose. I stood in front of the knot, which was about the height of my shoulder, and untied it, supporting the limb with my arm. As I began to retie it, Nacho flew over and landed on the limb, then sidled over to where I stood. "Silly bird," I said, my hands occupied.

Reaching over, he touched my cheek with his beak. I stopped. Slowly and carefully he traced the contours of my face, delicately, like a blind person. He gently grasped a lock of my hair, separated it into two strands, and let it go.

After I left the flight cage I stood briefly outside the clinic, overcome by something close to despair. People shoot crows for fun, I thought. How can I make them understand?

I took a few deep breaths. I had to check the birds in the clinic, feed and pack up the nestlings, and take the kids swimming. There was no time for this. You bonehead, said my practical side, get ahold of yourself. The damned crow could have poked your eye out.

At the pool we raced each other underwater, then the kids joined their friends while I watched. Neither complained about the nestling birds who had come to rule our summer lives, but I didn't know whether that was due to the adaptability of children or a buried resentment eventually sure to surface. I thought of Mac, perfectly still, Danielle drowsing contentedly on his lap; I thought of Skye, feeding nestlings, then speeding around the house for an hour to offset her ten minutes of suppressed activity.

My feelings of guilt were complicated. I was a working mother with no salary. I was always home but always busy. I worked days, nights, and weekends. John was supportive and interested, but was becoming increasingly exasperated with my unending workload. Was I asking too much of him? Was I giving my kids something valuable, or was I simply being selfish and catering to my own obsessions?

After dinner we all climbed onto the couch to watch a rented movie. Sometime later I felt Skye shaking me. "Mommy! Wake up! Have you been asleep this whole time?"

"I'm here," I said groggily. "I'm awake!"

"Do you know what's going on?" asked Mac, leaving me wondering how many ways he intended his question to be taken.

After the kids went to bed, I sat down with John. "I don't have an answer for you," he said. "What you do is amazing, and I think it's great for the kids—to

a point. Sometimes you're so stressed out that I worry about you. You don't ever get a break."

He rose and walked to the front door to turn off the light. "What's this?" he asked, looking through one of the small, narrow windows flanking the door. He stepped outside and picked up a cardboard box with six or eight small holes in the top and brought it inside.

"I rest my case," he said.

I pulled the note off the cardboard box. "My cat caught this bird," I read through clenched teeth. "Please help him."

It was a Baltimore oriole, a blaze of orange and black against a dirty white cloth. There was a dime-size hole in the base of his neck. One wing was shredded. His back was broken.

I put the box down gently on the table. "Dammit!" I cried. "They can't even knock on the door! They leave him and run!"

"Look at the handwriting," said John. "It's probably some teenaged girl."

"I don't care," I said viciously. "I hate her! I hate all these people and their goddamned cats!"

"I know," said John.

"Can you imagine what kind of pain he's in?" I said, blinking back tears. "I can't leave him like this."

I picked up the box, took a flashlight from the kitchen, stuck it in my back pocket, and went out the garage door, grabbing the shovel on the way. I walked into the moonlit woods, balanced the flashlight on a rotting log, and laid the gasping bird gently on a bed of leaves. "I'm so sorry," I whispered and brought the shovel down, my eyes streaming, hot fury burning through my stomach. Moonlight turned the forest floor silver. The flashlight illuminated a small, brilliant spot of orange.

"Are you okay?" said John, when I returned to the house.

"I'm fine," I said. "But I can't sleep. I'll be in later."

I filled a glass with vodka and ice and turned on the stereo. I flipped off the lights and closed my eyes, trying to think of nothing but the hoarse, rich sound of Herbie Mann's bass flute as it curled through the headphones, waiting for Cissy Houston's smoky, powerful voice to sing the blues.

Time flew toward the summer sky. The small spot of orange became a string of orange lights draped festively around my flight cage, shining into the darkness. The roof opened and fireworks shot straight up into the night and fell as birds, swooping upward before they reached the earth. The string of lights turned into a flock of orioles. And in place of the sound of explosives was a voice so beautiful it could ease a troubled mind and wash it all away. Like rain.

Chapter 34

UNDERSTANDING

"I'm calling you from my office," said the woman's voice. "We have a deck here, and there's a bird on it and he won't move."

"What kind of bird is it?" I asked.

"I don't know," she said. "It's brown. Maybe a pigeon or a wren or something."

Hmm, I thought grouchily. Maybe a kiwi or an ostrich or something. It's either a pelican or a spotted owl—hard to tell from this angle.

"Can somebody bring him to me?" I asked.

"Sorry," she said. "We're all working, and we can't leave."

"All right, I'll come over and get him. Can you put him in a box?"

"I'm certainly not going to touch it!"

"Is there anyone there who can put him in a box?"

"Nobody here will touch it."

"Oh, come on," I said, hoping to shame her into it. "Is there *a man* around there who can do it?"

"There's a man standing right here next to me," she replied, "and he won't touch it, either."

By the time I arrived someone had managed to get the bird into a box. Expecting a pigeon, I peeked in. It was a yellow-billed cuckoo, the first one I had seen up close. It was a beautiful bird, tall and graceful looking, with a white chest, brown back, boldly striped tail, and a slightly curved beak. Unless she had simply dropped from the sky onto the deck, it seemed logical to conclude that she had flown into the large sliding glass door.

I set her up in a reptarium in the shed, then readied a carrier for Danielle. It had taken some doing to find a home for the friendly mallard. I wanted her to have the company of both ducks and humans, but humans only up to a point: petting zoos were out. My meandering phone-call trail took more than the few days I had promised John, but it finally produced results. Sean Castellano, a rehabber an hour north of me who took in deer, raptors, assorted waterbirds, and miscellaneous stragglers, agreed to take Danielle.

"We have lots of ducks, quite a few mallards," he said. "There's always people around, so she can have company if she wants it. Otherwise she can hang out with the ducks."

The kids and I drove to Sean's. "I don't want her to go," said Mac, stroking Danielle's beak through the bars of her carrier. "I'm going to miss her."

"I know you will," I said sympathetically. "But living her whole life in our house is no life for a duck. She needs a big pond. She needs duck buddies."

"Look!" cried Skye. "Look at all the ducks!"

There were mallards, a pintail, and two mergansers; several large domestic ducks, a few Canada geese, a couple of greylags, and a swan. "There's a wood duck!" said Skye. "Just like Daisy but all grown up!" Some of them floated on the pond, some picked through the area near the barn, some drowsed in the sun. All looked quite content.

Sean came out of the barn and greeted us. "We'll keep an eye on her," he said to the kids. "Don't worry—she'll have a good home."

We walked down to the edge, Skye opened the carrier, and Mac lifted Danielle out and put her on the grass. She dabbled at their legs, then caught sight of the pond. Soon she was doing barrel rolls in the water, luxuriating in what she made clear was far superior to a bathtub or a plastic pool. She was chased briefly by one of the mallards, then one of the domestic ducks; she swam quickly away from each of them, and peace resumed. When Sean returned to the barn we gathered up the carrier and walked slowly toward the car, wondering if Danielle would follow us. Danielle, however, had no intention of leaving the water. When we finally reached the car she was floating tranquilly, not far from the wood duck.

"I hope you guys aren't too sad," I said on the way home.

"Some ice cream would help," said Skye.

The summer wound down. Had I been hired to create the sound track for a wildlife rehabilitator documentary, I would have recorded the wails and groans of exhausted rehabbers falling apart. "Aggghhhh," croaked Jayne when I called her to tell her I was sending someone with an injured woodpecker to her. "I swear to you, I'm not going to make it to the end of the summer. I had two volunteers scheduled for Saturday and neither one of them showed up and I had fortynine fledglings to feed. Halfway through the day I sat down and started crying. I swore I was going to quit, that this would be my last summer. But guess what? My chimney swifts are all flying! I throw them mealworms and they catch them in the air and then they just keep swooping around! You would die if you saw them—they are the most beautiful things in the entire world."

We chatted about our bird triumphs, our bird disasters, and our mental health. "Compared to you, I'm perfectly sane," said Jayne. "You're doing this with kids."

"I'm releasing all these birds," I said. "Sometimes I wonder if I should release myself."

"Forget it!" said Jayne. "You'll be out there having a great time and you'll hit a window or get attacked by a damned Cooper's hawk, and somebody will pick you up and put you in a box and bring you to me. And I'm too friggin' tired to take care of you."

I released the crows when the kids were at day camp and John was working, as Lo and Behold were skittish around other people and I didn't want them flying away in a panic. When I opened the door, the three of them flew into the trees and perched, heads swiveling, while I watched, willing myself to a state of calm. I thought of Lo jumping into the water dish, of Behold hopping up and down in place, of Nacho tracing my face with his beak. Something could happen and I would never see them again. Nacho coasted to the ground next to me, fluffed out all his feathers, and soaked up the sun, beak open, eyes half closed. Eventually he had enough, slapped his feathers down, and took off into the trees.

I lay back against the hill, cushioning my head with my arm. I should get up and make sure they're all right, I thought, just before I fell asleep.

◎　◎　◎　◎　◎

"Mom," said Mac. "I need a guitar."

"You *need* a guitar?" I asked.

"Yeah," he said. "I really do. I can't stop thinking about when I was playing Jack's guitar. I just . . . I just need to play a guitar."

Mac's cousin Jack had appeared at a family gathering with an acoustic guitar, playing several classical pieces and finishing up with "Stairway to Heaven." Mac had watched, mesmerized. Jack handed him the guitar and taught him some chords, and this was the result.

"Guitars are expensive," I said. "Let me talk to Dad about it."

"A guitar!" said John. "School is about to start, and he's going to be busy. He can wait for Christmas. It's nice that he wants a guitar, but there's a good chance that he's going to play it for a month and decide he's sick of it."

"He won't."

"How do you know?"

"Because he doesn't just want it, he *needs* it," I said.

A week later Mac and I were on our way to the music store, after I had placated Skye with a bag of new clothes. "I've gone out on a limb for you," I told him. "If you play this thing for a month and then quit, my credibility's out the window."

"Don't worry," he said. "I'm not going to quit."

It was an electric Fender, a good starter guitar that came packaged with its own amplifier. Mac quickly discovered how to use the Internet to learn songs, and for hours at a time I would hear notes and chords floating down from behind his closed door. "Come on!" I called. "The fledglings are all gone—I'm free! Let's get Daddy out of his office and go to Sunny Pond!"

"Let's go!" cried Skye, rushing down in her bathing suit.

"But I'm just in the middle of figuring out this chord," called Mac.

Sunny Pond was at the end of a long, hidden dirt road. It was filled with water weeds, and one had to keep swimming; if you paused for long enough near the edges, the odds were good you would emerge sporting several small leeches. But the leeches were easy to pull off and the pond was an oasis, sur-rounded by thick forest, tranquil and beautiful. Diving into the water, we all swam toward a jutting rock just large enough to accommodate us. I eased through the water using a slow breaststroke, watching the water striders fan out in front of me like a tiny herd of wildebeest on a liquid Serengeti.

"Wait!" cried Skye. "You want to play Not It?"

That night I called my aunt Sue Tyrie, who runs her own preschool, gives seminars on dealing with children of all ages, and has always been my child-care advice lifeline. She listened silently while I gave a garbled account of my last two summers, most of which was familiar territory to her.

" . . . and then we went to Sunny Pond and it was so nice and we haven't been there all summer, and I'm afraid I'm depriving the kids of all this stuff because I'm always busy and feeding babies or picking mealworms out of tanks or taking care of some horribly wounded bird. Do you think I'm a bad mom?" I finished, with a heavy sigh.

"How bad can you be if you're asking the question?" she replied.

"Pretty bad?" I asked.

"If this is what you love to do, I would not give it up," she said. "But I would find a way to cut down the number of birds you take in. The babies are the hard-est, right? Can you give them up for a couple of years, until the kids are older?"

"Well, I could . . . but where would they go?"

"They managed before you started, I guess they'll manage if you stop."

I filed this idea away, as I would not have to address it directly until the following spring.

◎　◎　◎　◎　◎

The kids started school. Each morning after they left I walked outside and greeted Nacho, who waited for his breakfast in one of the hemlocks near the parrot's outdoor flight. He would land near my feet, ready for a bite to eat, a game, and a head scratch; close by were Lo and Behold, always watching from above but with no intention of coming down until I had left the vicinity. None of the disasters I was sure awaited them had come to pass—at least not yet. With the summer over my bird population went down and I could spend time sitting alone on the rocky outcropping above the flight cage, notebook and binoculars in hand, watching as the crows explored, interacted with one another, and formed tentative connections with passing wild crows. It was blissful.

I cared for the other birds, greeted the kids when they came home from school, helped with homework, and wondered if any studies had been done on the stress level of wildlife rehabilitators. Late one afternoon I was sitting on the deck, staring at the sky and thinking about preparing dinner, when Tanya arrived with a skeletal great blue heron.

I could see no underlying cause for his emaciation, no broken bones, no open wounds. His rescuer had found him at the edge of a pond in a public park, barely able to stand. He could have been poisoned by lead or mercury, or had a heavy parasite load; but in the end it wouldn't matter unless I could bring him back from the brink of starvation.

In a perfect world I would have put him in the shed, filled up a rubber tub with live fish, and left him alone. But when I put him down he collapsed, too weak to raise his head.

Tri-State Bird Rescue in Delaware has developed an effective emaciation mix for waterbirds that can be pureed in a blender, then frozen. I pulled out a plastic container of the mix from the freezer, defrosted it, and calculated how much the heron could take, measuring out a little less just to be safe. Drawing it up into a large syringe, I attached a very long tube, then realized that John wouldn't be home for an hour.

"Mac!" I called. "Can you help me?"

You need four hands to tube a heron: one to hold the bird's neck straight, one to keep his beak open, one to thread the tube down his throat, one to push the plunger on the syringe, and one to make sure the tube doesn't come flying off the syringe when the liquid is ejected. Well, maybe five hands. Six, if you count the occasional straightening of the blindfold I always place around a heron's eyes so he can't see the monsters who are manhandling him.

Mac was unfazed. Wearing a large pair of ski goggles he calmly and quietly followed my whispered directions, reacting only with the occasional widening of his eyes. We finished and quickly left the darkened shed.

"Wow," said Mac. "What an amazing bird."

"We make a good team," I replied. "Can you help me again in a little while?"

John arrived home and reminded me that a babysitter was actually coming to our house after dinner, allowing us to join Alan and his wife Jan for a drink in a nearby town. "You know how hard it's been to schedule something with them," said John, when I protested that I couldn't leave the heron. "We're not even going for dinner! The heron should be settled by then, shouldn't he?"

As I readied another round of tubing mixture I drew up the full amount. Mac and I repeated our procedure, but before it had all been administered, a stream of the mixture ran from the side of the heron's beak. I gasped and pulled the tube out. I must have miscalculated the amount of liquid the heron could take, and it had backed up. Odds are he had aspirated some of it into his lungs, and would eventually develop pneumonia.

"What's the matter?" whispered Mac.

I used a small towel to soak up the remaining liquid, then we left the shed. "I made a pretty serious mistake," I said to Mac as we walked back to the house. "I tried to give him too much. I hope he'll be all right."

"He'll be okay," said Mac.

I said nothing further about it until John and I were en route to the restaurant. "It's my fault," I said, trying to keep my voice steady. "He'll probably get pneumonia and die, and it's all my fault. I'm so stupid!"

"It was a mistake," said John. "Everyone makes mistakes."

"Not as many as me. I've been making mistakes for two years."

"Of course, you have! The number of birds you've taken care of? If you hadn't made any mistakes you wouldn't be human!"

"But if I make a mistake, the bird can die!"

"If you did nothing, the bird would die anyway!"

We arrived at the restaurant and greeted Alan and Jan. "We've had a rough night," said John, and explained the preceding events.

"It happens," said Alan, as Jan nodded sympathetically. "Don't beat yourself up. If he comes down with pneumonia, you get him on antibiotics. Meanwhile, you did what you could."

"But it was so stupid," I said miserably. "And I left him in the dark—do you think I should have put a night light in with him?"

Jan looked at me in amazement. "Shut up!" she cried, grinning incredulously. "You think he's going to live or die depending on whether you used a night light? Oh, man—you're a mess!"

I stared at her, shocked, then started to laugh.

"I have to agree," said Alan, deadpan. "You *are* a mess."

"I've been telling her that for months," said John.

"You know what you need?" said Jan. "A lemon drop martini."

"All around," said Alan.

When we returned home I walked back to the shed, buoyed by my husband

and good friends. I found the heron lying motionless, his head resting on the floor and his wings extended, as if in flight.

The night was warm and filled with stars. I vowed to myself that I would take no more herons, knowing if I kept my promise I would never find redemption for the one I had lost. I wondered if anyone in my fragmented life could have predicted the direction I would take, then I remembered something I hadn't thought about in thirty years.

I was sixteen years old, in boarding school, and a battered horse van pulled up to the stable run by my friend Sam, the ex-jockey. The horses inside were cheap and anonymous, some broken down, some intractable, all offered for sale by a loud tough guy named Ray, a man who set my teeth on edge the moment he jumped down from the driver's seat. Sam wanted to see the horses in action, so I climbed onto the first one, a thin black mare trembling with fear. Responding to my encouragement she trotted a few steps, then without warning crashed to the ground. I rolled away and climbed to my feet in time to see Ray run toward her, red with anger, and deliver a heavy kick to her side. I rushed to where Ray stood and he turned toward me, his watery blue eyes wide with surprise.

"Leave her alone!" I screamed, and hit him in the face.

Ray let out a string of obscenities and raised his fist, and Sam appeared out of nowhere, hobbling up between us like a human wedge. "I'll buy her, I don't care what it costs," I shouted. "I'll kill her before I sell her to you," Ray shouted back, and Sam pushed me toward the barn. "I'll take care of it," he said. "Don't make this harder for me than you already have."

I waited outside my own horse's stall, listening to the two angry voices as they rose and fell in the distance. I leaned my head against the door, thinking of how many men like Ray I'd met, how many more I'd heard about in the supposedly genteel world of East Coast barns and show stables. When I looked up, Sam was limping toward me, the black mare following behind him. He handed me her lead rope, his face impassive.

"You owe me a hundred bucks," he said.

I tried to hold back my tears and failed, crying for the black mare, for the ones left waiting on the truck, for all the beaten and broken horses I couldn't save. Sam stood unmoving, his hard eyes filled with sympathy.

"You're never going to have an easy time of it," he said quietly.

I gazed up at the moon. If I looked closely enough, I could see the heron's silhouette.

Chapter 35

TURNING LEAVES

"I found him on the ground at the golf course," said the male voice. "He couldn't fly—he wasn't even trying—and his head was going back and forth. Can I bring him to you?"

When I opened the cardboard box the red-tailed hawk seizured, staggering backward and thrashing for what seemed like an eternity. When he finally quieted, I transferred him to a padded carrier and draped a towel over it.

"It could be any number of things," I said, handing the man a pen and paper. "I need your name, address, and phone number for my records, but I'd also like the name of the golf course. Some of them use a lot of poison, and that could be the problem."

"Could you call me when you find out?" he asked. "He's such a beautiful bird, and I'd just like to know."

I couldn't rule out that he had been hit by a car or suffered some other kind of trauma, but when I examined him a half hour later I could find no signs of it. He was alert and only slightly thin. I gave him fluids and was putting him back in the crate when he seizured again. In less than a minute he was dead.

I called the golf course and demanded to know what kinds of poison they used. I was transferred to the groundskeeper, who tried to be helpful.

"We're very careful about poisons," he said. "We don't use many at all, compared to some of the courses around here. We use Merit, Sevin, Telstar, and a few different fungicides."

Merit contains imidacloprid, a chlorinated nicotinoid compound that affects the nervous system and, according to the label, is "particularly toxic" to earthworms and bees; Sevin contains carbaryl, a cholinesterase inhibitor, and is "extremely toxic" to bees and fish; Telstar contains bifenthrin, which is "moderately toxic to many species of birds, toxic to bees, and very highly toxic to fish, crustaceans, and aquatic animals." This lethal cocktail was being poured all over an area frequented by wildlife—and this was the *careful* golf course. That was bad enough. But how did I know the golf course's snack bar wasn't using rodenticides, a far more likely cause of the redtail's death? I was livid. Determined that someone would pay for this beautiful hawk's terrible death, I called Shawn Rogan, who works for the Putnam County Health Department. Shawn responded with his usual swiftness.

"Are you going to be home this afternoon?" he asked. "I'll stop by and pick the hawk up, then I'll send him to Ward. Don't worry, we'll find out what happened."

The New York State wildlife pathologist is Ward Stone, who has worked for thirty years diagnosing and tracking causes of wildlife mortality, including chlorinated hydrocarbon insecticides (such as DDT) and organophosphate pesticides (such as diazinon, which was banned from use on golf courses and sod farms in 1987, after Ward testified as principal expert for the EPA). Although his budget is slashed with depressing regularity and he works with a skeleton crew, Ward and his staff have monitored outbreaks of botulism and the secondary poisoning of raptors by rodenticides, followed the deadly trail of PCBs through the Hudson Valley, and documented the initial outbreak of chronic wasting disease. Despite the obviously critical importance of this position, New York is one of the few states that have an official wildlife pathologist.

I had known Ward for years, and always marveled at the way he managed to combine his passion for his job with being a father to six kids—who, unsurprisingly, were always jumping out of windows and dragging things into the house from the woods. "How are your kids?" I asked, when I called him two weeks later.

"They're so much fun," said Ward. "They're such great kids. I was just teaching Jeremiah to fly-fish the other day."

"Where did you go?" I asked, envisioning one of the paradisiacal upstate trout streams.

"Well," he said, "actually, we were in the living room. I didn't mean to do it, it's just that he asked the question and I wanted to answer him before we were distracted. Speaking of which, your redtail was positive for West Nile."

I was shocked, even though I knew about the devastating effects the mosquito-borne West Nile virus (WNV) was having on bird populations across the country, especially on raptors and crows. My second trusty electronic mailing list was Raptorcare, a list devoted solely to raptors and moderated by Louise Shimmel, the director of the Cascades Raptor Center in Eugene, Oregon. All summer the list had buzzed with desperate rehabbers pooling information, trying to come up with a protocol for a disease that was not yet fully understood—in birds or in humans—and had no proven treatment. Doctors treating West Nile in humans were contacted; the information from those willing to discuss possible raptor treatments posted; contact numbers for drug representatives able to give discounts to rehabbers were exchanged. List members wrote painstakingly detailed summaries of treatments that seemed to be working, hoping their observations might help a rehabber in another part of the country save a stricken bird.

"It is like they are seeing demons," wrote Marge Gibson, who founded and runs the Raptor Education Group in Antigo, Wisconsin, and who created a West Nile informational database when the virus first appeared. "Hallucinating. We need to think brain insult, with neuro pathways having to regrow and reconnect. All you can do is wait and see what happens next, then watch and adapt as you treat the symptoms. WNV birds are some of the most emotionally and physically draining patients you can have. Remember to take care of yourselves."

During the summer I printed out all the West Nile information that had come in on the electronic mailing list, and I collected it into a separate gray three-ring binder, hoping I wouldn't be called upon to use it. I read heart-wrenching e-mails from raptor people in the Midwest who had received dozens of wild hawks and owls with West Nile, who had spent days wrapping their flight cages in mosquito netting only to watch several of their beloved longtime education birds die in convulsions. Although I had read that several dead birds collected by the public from areas nearby tested positive for the disease, until now it had all been slightly unreal. Until now.

"If I get other birds and suspect West Nile, should I send them to you?" I asked Ward.

"Absolutely," he replied. "We need all the information we can get."

⊙　⊙　⊙　⊙　⊙

During the next ten days I took in two crows. One died ten minutes later, the other lasted a half hour. Shawn sent them up to Ward, and both tested positive for West Nile. I braced myself for what I was sure would be a flood of West Nile birds, only to have the disease disappear—for the moment—from my doorstep.

I resumed my long-neglected daily run. Spooked by West Nile I traversed the woods, searching for the goshawks whose young had left the nest months before. Within a few days, they found me. As I ran toward a high overlook I heard a strange little grunting noise, a bit like a squirrel but not like a squirrel, which seemed to follow me up the trail. I stopped. It was the male goshawk, and as soon as I spotted him I heard, floating through the trees, the clarion call of his mate. "Uh-oh!" I called to him. "Sounds like your wife is on the warpath again!" Moments later she blazed into sight, screamed a few epithets at me, and flew off. I spent several minutes talking to the male, flattering him and cooing at him, and when I finally continued on my way he came along for the ride.

We headed up the trail, I running, he flying from branch to branch, until we came to a small clearing bathed in morning sun. He settled on a thick limb, I on the mossy ground below. Fluffing his feathers and shifting his weight, he kept one eye on me and the other on his surroundings, while I leaned back against a huge old oak tree. Before long he lifted one foot, a sign of comfort and peace.

I tried to live in the moment, to think of nothing but the deep blue of the sky, the warmth of the late September air, and the goshawk perched, miraculously, above my head. But I couldn't shut out the world at the edge of the woods, where a deadly disease that originated thousands of miles away could claim the lives of both goshawks and I would never know it. Once again, I could neither protect nor save them.

Here be dragons.

But I still had the ability to turn away from an overwhelming reality. I forced the dark thoughts aside and raised my eyes, wondering if the wild hawk who had given me the gift of his company could sense my gratitude. Finally I rose and continued my run, heading for the top of the ridge. From time to time I'd see the goshawk flying through the trees, sometimes ahead of me, sometimes to the side. I reached the summit and noticed two quick shadows pass over my own. Looking up, I saw that the male's mate had joined him. Flying wingtip to wingtip, they vanished into the canyon.

⊙　⊙　⊙　⊙　⊙

"Mom," said Mac.

I was sitting at my computer, printing out a long e-mail message from Linda Hufford, an Austin, Texas, rehabilitator who knew far more than I did about caring for great blue herons and was happy to share her expertise. I looked up from my computer. He and Skye were standing side by side, looking like a pair of sad-eyed waifs from a Keane painting.

"Go ahead," I sighed, familiar with the look. "What do you want?"

"A dog," said Skye.

It was not a new request. The dog idea had started the previous fall and continued sporadically through the summer, during which any mention of another creature to care for had sent me into a wild-eyed frenzy.

"A dog," I said slowly, as if I had never heard the word before. "Let me ask you something. If you had the choice of getting a dog or going to Disney World, which would you choose?"

John and I had discussed the dog idea. Although I wasn't eager for a dog, I honestly believed the kids should have one; John was simply not eager for a dog, period. Both of us thought it should be put off. The plan had been: (1) avoidance, (2) distraction, (3) bribery. We had already covered the first two; Disney World was the bribe. Since we had not yet taken the kids there, it seemed to be a perfect way to combine plans one and three. It was not something offered lightly, either: if I died and went to hell, I would find myself either in a shopping mall or at Disney World.

"A dog!" they chorused, loudly and without hesitation.

"Hmm," I said. "Let me think about this."

John and I had a summit. "If I give up doing baby songbirds, we could have halfway normal lives and the kids could have a dog," I said.

"With one hand she giveth, with the other she taketh away," said John.

The kids were delirious. I was a confirmed stray dog owner and had never bought a dog in my life, but suddenly I wanted to know exactly what we were getting. We hit the Internet in quest of the perfect dog, and after hours of searching, a photo popped up on the screen of what looked like a Labrador with poodle hair. There was a moment of silence, then a rush of approval from—astonishingly enough—both kids. " 'Curly-coated retriever,' " I read. "That's a nice-looking dog! Maybe we should . . . hey, wait a minute! They're bird dogs!"

As it turned out, curly-coated retrievers had been bred since the 1700s to retrieve the wild birds their owners slaughtered in the field. "Guys," I said, "this could be a deal-breaker. Think about it—what happens if I finally release a bird and the dog brings it back?"

We went to a New Jersey dog show, met two curly-coated retrievers, and were assured by their owners that a puppy who grew up around birds could live in harmony with them. Exchanging looks of trepidation, John and I signed on for a puppy. "Now we just have to wait for the phone call," I told the kids. "We're last on the list, and the mom might not have enough puppies for us to get one. Meanwhile, I'll have to contact some other breeders. Best case—it'll be a few months."

"Ohhhhh," they sighed mournfully. "What will we do until then?"

With few birds to care for we took day trips, went to the movies, and at the end of October, all drove down to the Bronx Zoo. We spent the day wandering through various exhibits and habitats, saving the enormous World of Birds

building for last. We entered a warm and humid jungle, a bird lover's dream, lingered along beautifully photographed informational areas, and finally found ourselves in a darkened hallway. The exhibits before us were like living dioramas, habitats filled with light and live plants and flowing water, separated from human spectators by a single railing and the fact that birds won't willingly leave the light and fly into the dark. It was a weekday afternoon, the crowds were sparse, and when we reached the North American Forests exhibit the four of us were the only spectators. We spotted a robin, a junco, two goldfinches. They weren't the one we were looking for, however.

"There he is!" cried Mac.

At the end of the summer I had only one unreleasable bird: the friendly little nuthatch, raised in captivity after being found as a nestling with a broken wing, whose fracture had not healed well enough for him to live in the wild. Nosy and energetic, he had always greeted us by creeping along the mesh-covered ceiling of the flight cage upside down, then stopping and emitting a cheerful *"henk henk!"* A greeting we always returned.

He was in the back, investigating a knothole in a pine tree. He looked up, flew toward us, and landed on a tree in the very front of the exhibit, perhaps five inches from where I stood. He peered up at me. "Honey!" I said. "How are you?"

He crept up and down the tree, chattering, and when we walked to the other side of the exhibit he followed us, landing on a nearby limb. As a small crowd of people entered the room, he flew to a far corner and watched them from the safety of the pine tree. As soon as they left he returned, hiking jauntily across the rough bark, having picked four familiar faces out of the river of humanity constantly flowing by.

"Henk, henk," he said.

"Henk, henk," we all replied.

Chapter 36

THE QUIET SEASON

Hawks and owls (as well as most other kinds of birds) are frequently hit by cars. Waiting in a tree, raptors will spot their quarry and immediately swoop across a road to catch it, having never been told by their parents to look both ways before crossing. This beautiful young redtail had been lucky; he had received a glancing blow instead of the full impact. His left wing drooped but was not broken. A bad bruise often takes longer to heal than a fracture, and since I wasn't going to release a juvenile redtail in the middle of winter, I decided to let him ride it out with me.

It was late November, I had few birds, and the redtail turned out to be a mellow and easy patient. During the ten days he spent resting in the shed I followed the advice of the pros on Raptorcare and turned my songbird flight cage into a raptor overwintering enclosure. Cutting a huge, dark brown oilcloth into strips three inches wide and ten feet long, I hammered them around the sides of the flight in three-inch intervals, hoping they would look solid enough to prevent the hawk from trying to fly through the mesh-covered hardware cloth. I covered the big permanent perches with new Astroturf, carried in a horse-size rubber water dish, dragged in several large cut logs, and tarped part of the roof. Luckily all the work paid off; when the redtail finally moved out to the flight cage, still favoring the wing but much improved, he settled in with no problem. I called Paul Kupchok.

"Paul!" I said. "Do you have a redtail I could borrow? I have a hit by car in my flight and I don't want him to spend the winter alone."

"Sure," said Paul. "You want some kestrels, while you're at it?"

A slight problem arose: freezer space. I had been using the bottom drawer in our upright garage freezer for raptor food, as it was fairly hidden. But when I added bags of frozen fish, plastic containers of emaciation mix, and the occasional carefully wrapped songbird who hadn't made it but who could be used to feed an ailing accipiter, things started to get tight. Tight as well as awkward.

"No way!" Skye would shout. "I'm not opening that freezer! If you want a loaf of bread you have to get it yourself!"

Realizing that keeping two large redtails through the winter would entail stocking up on an even greater number of dicey items, I went through the Flyaway bank account and found that I had enough to buy a small chest freezer. If I rearranged the garage I could nestle it in right next to our upright, and the food items of birds and humans could exist in separate-but-equal harmony. As I wandered through the freezer section of a local appliance store, a man swaggered up to me.

"So!" he said, wearing the expression of a guy used to charming the ladies. "Pizza or venison?"

"Rats," I said. Like the Cheshire Cat in reverse, his smile vanished first.

I called Rodney Dow. Rodney is a model of self-sufficiency; if hostile forces took over our town, Rodney's family could hunker down for years while Rodney kept the enemy at bay. He grows vegetables, keeps bees, stocks his pond with fish, built his own smokehouse, and in the fall, hunts deer, stocking his freezer and eventually using every part of the animal. Although I am firmly anti-hunting, I make a grudging exception for deer, as thanks to humans there are far too many for the land to support and starving to death is a bad way to go. Last fall Rodney had called and offered various cuts of venison for my raptors, but I hadn't the room to take him up on it.

"You bet," said Rodney. "I'll bring you as much as you need. And I have some mice for you, too, that I trapped in the smokehouse."

The day after we celebrated Winter Solstice a car pulled into the driveway, and a man emerged dressed in a festive Mexican vest and a fuzzy Santa hat. It was Frank Olivetto, a former IBM exec who, for a period during his semi-retirement, worked at the small local country club nearby. He fed the ducks that took up summer residence on the club pond, gave holy hell to any golfers who aimed at the visiting Canada geese, and gave me detailed reports on the comings and goings of whatever birds happen to be passing by.

"Merry Christmas!" he announced. "I have a present for your hawks!"

He pulled out a cooler filled with venison. "There's meat, organs, bones, there's everything they'd want," he said. "Just tell them it's from Uncle Frank."

The stress of the summer began to dissipate. I had the two redtails and a couple of songbirds, and things were manageable. By the end of a hectic summer season most rehabbers have little goodwill toward men; we have little goodwill toward anything human, the source of most of the damage we're trying to undo. But I was touched and inspired by people like Rodney and Frank; and as I looked back over the summer I remembered Ed's words and thought of the people who had spent a great deal of time and effort to find help for an injured bird.

I saw the crows every morning, but as the weather grew colder I saw them less often. Nacho, who had always flown down to greet me when I appeared with his breakfast, would no longer let me touch him; he flew along with me to where I placed the food, but then stayed in the trees with Lo and Behold until I left. He still chattered to me, sometimes fluffing out his neck feathers, bobbing his head up and down and purring "*ooh, ooh, ooh*," which I would happily mimic back to him. Sometimes I would see the three of them fly by with two other crows, then all five would disappear. But for the small plastic container of crow food I left for them each morning they were wild crows, and although I missed my interactions with Nacho, I rejoiced in their freedom.

As Christmas approached there was a faint background sound to Skye's beloved Christmas carols: chord progressions, rolling off Mac's guitar from behind his closed door. Practicing for hours on end, his unruly blond hair reaching halfway down his back, he seemed every inch the budding rock star. He had found his passion, something that could help him weather the highs and lows of his childhood and early adolescence. I wanted the same thing for his sister. But what would work for Skye, who was visual, tactile, yet unable to sit still unless she was in front of a computer screen?

On Christmas Day she opened one of her boxes and stopped dead. "A digital camera," she whispered.

By that night she had figured out all the camera's tricks and was downloading her first batch of photos onto the computer, feats that would have taken me months to duplicate. Before long she found her signature subject: the slightly odd still life. Arranging a flower or a series of beads, or peering down the neck of a plastic soda bottle, she would illuminate her subject with a lamp or flashlight and shoot it from a strange angle, making it recognizable only after a second glance. Eventually she started shooting moody outdoor pictures: a ghostly moving swing, the hurried swirl of a flock of sheep, a bridge through a rain-drenched car window. Snow fell and the temperature plummeted, but John and I were filled with a sense of well-being: the kids were happy and occupied, and all was right with the world.

By the end of January, however, there was a thick cover of snow and for weeks the temperature hovered between ten and twenty degrees Fahrenheit. The crows arrived each morning and had their breakfast, the resident songbirds stayed by our well-stocked feeders, and the two redtails remained warm with generous helpings of rodents and venison, but I knew that most of the wild creatures were hungry. One morning I glanced out at the blue jay feeding platform I'd built next to the house and sitting there, fluffed and still, was a black vulture. I had never released a rehabbed black vulture here, nor had I ever seen a wild one nearby. The vulture sat quietly, no more than twenty feet away, watching the house.

"He's not one of ours," said Mac. "I wonder why he's alone?"

"Quick," said Skye. "We have to get him something to eat!"

Luckily I hadn't yet fed the two redtails, so I hurried outside with one of their rats. After slicing it lengthwise (the vulture would have had a hard time getting through the thick skin), I left it on the driveway just outside the garage, and before I'd even made it back to the door he was eating.

The vulture set up camp in a huge old tulip tree. The tree's lowest limb, a massive arm twenty feet up, hung over the driveway; in the morning he'd sit on the limb, waiting for his breakfast, and if we went out at night our headlights would briefly illuminate his solitary silhouette. I left his food not far from the tree, near the short trail that led to the wide, barberry-choked field. I could watch the area from my kitchen window and always kept the binoculars handy for a closer look. One day as I was delivering the morning meal I heard a familiar voice.

"There's a goshawk out there!" I told John later. "It's a young one. I can hear him calling!"

"Oh, great," said John. "Is he calling my name?"

Two days later I looked out the kitchen window and the large bird in the middle of a morning feast was not the vulture. I grabbed the binoculars and focused on a young goshawk, its brown and white juvenile plumage so streaked and speckled that it looked almost checkerboard. It ate hungrily, constantly surveying its surroundings with fierce yellow eyes, eyes which would eventually turn orange and finally red at maturity. From then on each morning I put out food for two, silently thanking Frank and Rodney. Occasionally I would see the hawk and the vulture eating at the same time, eight or ten feet apart.

They stayed until the temperature rose and the snow began to melt. One day the vulture showed up with four others, and after a few hours spent perching around the house, they all left together. The goshawk stayed a week or two longer, coming to eat every few days instead of every day, then he too disappeared. The first year of a young raptor's life is hard, and 80 percent do not survive. I hope the extra food helped him make it into that lucky 20 percent. For days after he stopped coming I could hear his voice in the woods, an echo of summers past.

"I know your parents," I wish I could have told him. "Would you send them my regards?"

◎   ◎   ◎   ◎   ◎

We were bumped from puppy list after puppy list. We were at the tail end of each one, as we weren't going to show our dog and didn't want him for hunting. But then in February we heard from Kathy and Scott Shifflett in Maryland, who said that one of the prospective puppy owners had dropped from the list and we were in. Soon we received a photo of the proud mom snuggled next to a group of small fuzzy shapes, a carefully drawn arrow pointing to the one that was ours. I could drive down to pick him up in mid-March.

By the time we received the phone call the excitement had reached a crescendo. "I'm really sorry," said Kathy. "But he's lame. He's growing so fast his joints can't keep up with him. I know you were looking forward to getting him next week, but I don't want him to leave here until he's perfectly sound. The vet said he should be fine in another month."

"All right," I said. "Thank you. I appreciate your concern for him. But how big do you think he's going to get?"

"Big," she said. "They're usually about eighty pounds, but this one is going to be big. He could reach ninety."

"Ninety!" I said, my heart sinking. Little did either of us suspect that ninety pounds would be but a brief pause on his way to maturity.

On a chilly April morning I loaded a puppy crate into the back of my car and headed south to meet Scott Schifflet. I pulled into a highway rest stop in Delaware, searching for a man in a blue jacket walking a small black puppy on a leash. What I found was a man in a blue jacket walking a black puppy the size of a full-grown cocker spaniel. I parked and approached the duo hesitantly. The puppy had a curly coat.

"Suzie?" called Scott.

Scott and I shook hands while the oversized creature danced awkwardly around us, exuding forthright puppy goodwill. When I knelt down the puppy furiously wagged his tail—which was as long as the rest of him—and gazed up at me with gentle, slightly hooded brown eyes. In an instant I decided that for all I cared, he could be the size of a camel. I tucked him carefully into the crate, and he slept almost the whole way home.

The winner of our Name the Puppy Contest was Merlin, straight from the world of magic, fairies, and dragon-riding heroes, and coincidently, also a kind of falcon. For the kids it was love at first sight, and John quickly fell under his spell. The only holdouts were the parrots. Mario glared at him from various high perches and refused to go near him; Zack, naturally, preferred the hands-on approach. Sitting quietly on a chair arm, he waited until Merlin ambled over; when the puppy stuck out a friendly nose, Zack sank his beak into it. Merlin howled and stumbled backward, and the kids rushed to his rescue. Zack spent the next half hour laughing uncontrollably from his cage, where he was given a time-out. As it turned out, curlies *are* intelligent: from then on Merlin made a wide circle around any and all birds, and never gave them a moment's trouble.

As spring arrived I took stock. Things could be a lot worse, I concluded. And if I can just stop taking nestling songbirds, things could be a lot better.

PART THREE

* * * * *

Chapter 37

UNRAVELING

The springtime birds began to trickle in. I brought the two redtails who had spent the winter in my flight to Paul's, where they spent a few weeks catching live mice in his far bigger flight in preparation for their release. Since they were young and had not yet established their own territory, Paul decided to release them both in the nature preserve adjoining the Green Chimneys School. A small group of us watched as the hawks were tossed into the air, one by one, and with clean, powerful strokes flew upward and away. I watched them go with exhilaration and regret. I had spent months appreciating the varied facets of their personalities, and the odds were I would never see them again; yet they were going home, back to where they were supposed to be.

I recorded a message on my phone machine saying I no longer took young songbirds, envisioning that the ensuing gaps in my schedule would be filled by the kids and puppy. Instantly the calls for every other kind of bird seemed to double. I juggled furiously, passing the names of other rehabbers on to callers, trying to figure out ways to streamline what was essentially an unstreamline-able process. My facilities weren't extensive enough to have separate, perma-nent areas for songbirds, raptors, and waterbirds, so when a new one arrived I usually had to struggle to shift crates and enclosures around. The logical thing for me to do would have been to narrow my bird focus yet again, but I had just abandoned the area's nestling songbirds and I was too guilt-ridden to turn my back on yet another segment of the avian population. Instead I made do, sometimes ignoring the phone for several hours, hoping by the time I returned the call they would have found someone else, and at least able to say that I was no longer bound to a thirty-minute schedule.

The goshawks left the nest that had given them so much trouble and moved deeper into the woods. We could hardly remember a time before Merlin, who was so good-natured he wagged his tail in his sleep. My life balanced precariously between birds and family, with little time for anything else. "Don't you ever take time off?" asked our friends, whom we seldom saw except in passing.

"We have a goose here," said Robin, calling from the Cortlandt Animal Hospital. "It's a . . . hold on . . . let me ask—what kind of goose is it again? It's an American brant. A woman named . . . let's see, Amanda Sewandowski . . . was driving down the road and saw this goose standing on the shoulder, and the guy in the car in front of her actually swerved so he could hit it. The poor woman—she slammed on her brakes in the middle of the road and jumped out of her car so she could rescue the goose. She said she was sure they were both going to end up like pancakes."

The brant was a beautiful dark bird with a white chinstrap and white around her tail. I blocked off one corner of the shed with a metal dog pen, settled her on a towel-covered section of an egg-crate foam mattress, then hurried to my computer. "Help," I wrote to my wildlife rehab electronic mailing list. "I don't know anything about brants!"

Within two hours there was an answer. "Hey, it's Wendi!" came the cheerful reply. "I can help you with your brant. I'm in Montana—call me!"

Wendi Schendel is my Internet friend who used to raise wood ducks and gave me all kinds of advice when we were caring for Daisy. As it turned out, she was also an expert on sea ducks and geese and had worked with many brants.

"She was on the side of the road and some savage deliberately ran her down," I told her. "Nothing broken, but she's emaciated and really badly bruised."

"Poor thing!" said Wendi. "Sounds like she was migrating and got all skinny and exhausted. Sometimes they get disoriented—the roads can look like rivers if there's fog or the roads are icy. I'd say tube her for a couple of days, then get her on trout pellets and greens. These guys are not easy—they're really nervous. They tend to panic. Put up some curtains for her so she feels more secure. And call me back so you can tell me how she's doing!"

I draped dark blankets all around the brant's pen, hoping it would make her feel safer. She was scared to death of me, and every move she made was painful; immediately after she startled I could see her flinch. I tried never to look at her directly, and as often as possible turned my back to her. Trying to be solicitous of her frayed nerves, I avoided any sudden motions and instead moved slowly and fluidly, as if I were underwater. "The brant is teaching me Tai Chi," I told John.

"Maybe I should do that with you," said John.

The recovering goose responded, calming down and eating on her own. Although I was gratified by her progress, I began to awaken in the middle of each night and stare at the ceiling, unable to return to sleep. There were too

many things to think about: the birds who were going downhill, the ones who might not make it until morning, the ones who had suffered from the mistakes I had made. I lay awake thinking of the vulture found by the side of the train tracks, clinging to life despite broken bones and the gaping hole in his chest. The young peregrine hit by an airplane, his lower wing shredded, the jagged bones exposed. The cormorant with fishing line encircling and imbedded in his legs, the hook lodged deep in his abdomen.

I tried to think of the man who had telephoned me and stayed by the vulture until I arrived. The people at the airport who found the peregrine and then spent more than an hour on the phone, looking for help. The three guys out in a fishing boat on the Hudson, who had cut their trip short when they found the cormorant so they could hurry him to the vet.

I tried to think of them, but my mind veered away, focusing instead on the teenager who deliberately ran a Canada goose down with his jet ski. The developer who threatened to sue if he couldn't destroy a pristine wetland. The latest assault on the Endangered Species Act. *How can wildlife survive?* I thought. *How do any of them stand a chance?*

◎　◎　◎　◎　◎

During the next three years Merlin grew into a 115-pound dog with a deep bark and a fondness for footwear. He always believed he was hungry, even if he had just eaten, and occasionally he would lose control and scarf down one of the kids' socks. After the first panic-stricken rush to the vet I started stocking up on cat hairball remedy, an entire tube of which could usually grease Merlin's insides enough to allow him to pass the sock, the very idea of which would send the kids into paroxysms of dismay. He had a bullwhip of a tail that could clear a coffee table, inflict an actual bruise, or tow a child all the way around Sunny Pond. He was the best-natured dog any of us had ever encountered, and we never stopped feeling lucky to have him as part of our family.

After several guitar teachers had to abandon their lessons in favor of other work, Mac found Jake Harms, a fourteen-year-old prodigy who ratcheted up Mac's already formidable enthusiasm several notches. Mac found a bass player and a drummer and formed a band, playing White Stripes's "Seven Nation Army" for their first gig, the school talent show. Mac aced his solo—at least, the audience thought he did—and the band received a standing ovation. But later that night I found him in his room, slumped in his chair.

"I didn't ace my solo," he said. "Didn't you hear the second verse? I messed it up."

"Are you kidding?" I cried. "I've been listening to you practice for weeks and I didn't hear anything wrong!"

He stared miserably out the window. "I can't play like Jimmy Page," he said.

"Mac," I said, pulling up a chair. "You're eleven years old. Jimmy Page couldn't play like Jimmy Page when he was eleven. Give yourself a chance."

"I messed it up," he said.

"Don't do this to yourself," I said. "You're a really, really good guitar player."

Skye held on to Marigoldy, even after she figured out that her mother was actually writing the notes. I presented her with a shoe box filled with the letters she had written to various fairies, which she placed on her shelf next to her own carefully hoarded box of fairy responses. Every once in a while she would sigh wistfully and say, "I wish Marigoldy would write me another note," and in the morning it would appear, tucked beneath her pillow.

She took several local digital photography classes and learned how to manipulate color and shading, then joined a monitored Internet photography club where young photographers shared their pictures and exchanged comments. She carried her camera everywhere, pulling it out when she found a striking image or when a new situation made her feel awkward. Trying to find our way to the Brooklyn Museum one Sunday, we gave up deciphering the subway map and asked a young woman how to get there. "We're going to see the Annie Leibovitz exhibit," said Skye.

"So am I!" said the woman.

"You're not a photographer by any chance, are you?" I asked.

"Sure am," she replied.

Soon we were sharing a subway seat and chatting about cameras, photography courses, and what it was like to be a professional photographer. "Let's move to New York City!" said Skye later.

During the day I was thrilled by their interests. But in the middle of the night, sandwiched between thoughts of the day's patients, came pinpricks of realization: my kids were slowly starting to move away from me. Had they been songbirds, they would have been determinedly hopping in and out of the nest, wanting to explore, to make their own decisions. These were all milestones to celebrate. But I knew what came next.

John began a new book. He traveled, interviewed people, and attended seminars and conferences, returning relaxed and invigorated, filled with stories and new ideas. He was never sure what he was returning to, however. Sometimes he would find me celebratory, after a bird I was sure was a goner had essentially risen from the dead. Other times he would find me edgy and knotted, when the losses that I tried to force from my mind kept rising to the surface.

"That new crow has pox," I said to him once, distraught, sure that she had somehow infected my four fledglings. John nodded and started to leave the room.

"Thanks a lot for the sympathy," I said sarcastically.

"Sympathy!" he said. "I'm sick of giving you sympathy! You're always upset about something, there's always some bird crisis going on! How much more sympathy do you want?"

I had long talks with Wendi Schendel, who not only helped me through rehabbing the brant but found a sanctuary where she could be with other geese before she was released. "You're too isolated," said Wendi. "You don't have any downtime and you don't have any help. Believe me, I know—I went through this myself. You need to specialize; you can't keep taking all these different species, given the way you're set up."

"I know," I said. "It's just so hard. People call me on the phone, they appear at my house. . . ."

"You have to tell some of them no," said Wendi.

"I can't!" I said.

"Then you need to shut down for a while," said Wendi firmly. "You're burning out."

At least I could still mock myself. After a particularly trying day I would grab a beer, put a CD on the stereo, and with Mario perched on my hand, lean against the counter in the kitchen. As the mournful, three-note bass riff began, Mario and I listened to R. L. Burnside murmur his weary way through "Bad Luck City." We shifted from side to side, slowly nodding our heads to the music, as if we'd both been ridin' that downbound train way too long.

When I sliced defrosted rodents into tiny pieces or cleaned out the mealworm tank I occasionally thought about what I would be doing had I been a more obedient daughter: relaxing on the yacht club porch, wearing crisp white linen, and drinking iced Southsides. I tried to juxtapose the two lives and envisioned myself peering through designer eyeglasses at a letter:

Dear Ms. Gilbert,

On Thursday, July 21, one of our members saw you scraping a dead squirrel off Harbor Way. According to the member, when a car slowed down you shouted, "I'm hungry! Do you mind?" This is hardly the first complaint we have received about you. We are not amused.

Please appear on August 1st for a membership review meeting.

Sincerely yours,
The Membership Committee

I continued to try to pass birds along to others. I drove orphaned raptors to the Raptor Trust, in Millington, New Jersey, a renowned wild bird rehabilitation center founded and run by Len and Diane Soucy. Advocates for birds of prey for thirty years, the Soucys have expanded their original backyard operation into a center with a state-of-the-art hospital, an education building, an enviable collection of outdoor raptor housing, and paid staff members who are compassionate and skilled. They have surrogate parents of many species, allowing orphans to grow up under the tutelage of adult raptors instead of humans.

The Raptor Trust is ninety minutes in one direction, Jayne is ninety minutes in the other. I sent a few orphans and some adult songbirds to Jayne, who was always willing to take them as long as they were delivered and fit within her species boundaries.

"Jayne," I said. "I'm drowning. I can't keep this up."

"You have to start specializing," said Jayne. "Just don't specialize in those damned Cooper's hawks."

But those damned Cooper's hawks were exactly the ones I was contemplating, as well as their redtailed and sharp-shinned cousins, and their owl and vulture compatriots. I needed to draw a clear line: taking adult songbirds but not young ones was a line too easily crossed. Although I didn't have the enormous flight cages necessary to condition large raptors, I could get them back on their feet and then transfer them to Paul or to the Raptor Trust. I didn't receive nearly as many raptors as I did other kinds of birds. Young and injured raptors didn't need to be fed as often as songbirds. Maybe, at this point, it all came down to numbers and metabolism.

In September the kids went back to school and I searched for a permanent home for the summer's unreleasable songbirds. I had a wood thrush, a great-crested flycatcher, a chimney swift, and a diamond dove, a domestic species that had been found wandering along the sidewalk of a nearby town. I spent a few days calling around and finally reached the curator of the National Aviary in Pittsburgh.

"Sure, we'll take them all," he said. "Can you ship them to me by air?"

"I can't," I replied. "They're so little and delicate—I'd be a nervous wreck. Let me see what I can figure out."

The next morning I received a call about a red-tailed hawk trapped in a warehouse. "Lew!" I said into the phone. "Can you come with me? A redtail should be easy—we'll just chase him back and forth a couple of times and when he gets tired and lands, I'll grab him. They said he's been in there since yesterday."

The redtail turned out to be a Cooper's hawk, the bane of Jayne's existence, which made things many times more difficult. Smaller, quicker and more agile than a redtail, a Cooper's is difficult to catch under the best of circumstances, and the way the warehouse was built made it nearly impossible. Rows of shelves

and a large loft provided hiding places. Normally birds will remain immobile if the lights are suddenly turned off, but windows ran along one whole side of the building just under the ceiling, providing permanent light and convincing the hawk that they were her only means of escape. Refusing to go near the open doors or the truck entrance, she flew back and forth along the windows, hitting against them when we tried to drive her away.

I watched her fly swiftly the length of the warehouse, frightened but so far uninjured. If we can catch her, I thought, we might be able to release her right away—a bad situation with a feel-good ending. Maybe this is what I'm looking for. A sign.

The employees were helpful and concerned, watching as Lew and I failed to corner her with our long-handled fishing nets. When we tried to herd her toward the exits, she disappeared into the loft. I climbed the stairs and waited outside the closed door; when Lew shouted I inched through, spotted her on the floor, and threw a blanket over her. As I gathered up the blanket she somehow lunged out from under it, launched herself into the air and flew back toward the windows.

"She's got to be getting tired," said Lew. "Let's try one more time."

I stood beneath the bank of windows, hoping that when Lew shooed her I could lift my net up into the air and catch her. She flew toward me, but without warning she veered off and slammed into the glass. She grasped the window frame briefly, her neck at an odd angle. I dropped my net, rushed forward and caught her as she fell, sank to my knees so I could cradle her, watching as her breathing grew shallow and the light faded from her eyes.

"Please don't die," I whispered to her. "Please don't die."

On the trip home we were silent. Lew pulled into his driveway and turned the truck off, then turned and looked at me steadily. "That was a damned shame," he said. "Now, listen. We're going to get one of those mist nets, the kind they use to catch birds when they're banding them. We're going to stretch it between two poles so we can raise it up and the bird will fly right into it. And then this will never happen again. Okay? You all right?"

That night I dreamed it snowed in the warehouse, covering the desks and gathering in gentle drifts by the shelves. I followed the sound of wingbeats and found the Cooper's hawk crouching in the loft, her orange eyes burning into mine. I picked her up and she lunged slowly out of my hands, as if her escape had been captured on film and was being replayed inch by inch, frame by frame. *"Please don't die,"* I whispered to her, and as she veered away from the sound of my voice she slammed into the window and shattered like glass, each jagged piece of her falling, one by one, into the snow.

"I have an idea," I said to the kids. "Let's take a road trip to Pittsburgh! If we drive all those unreleasable birdies to the National Aviary, the director said he'd give us a private tour—at feeding time! Wouldn't that be cool?"

"Pittsburgh!" said Mac. "How far away is Pittsburgh?"

"Well, it's a haul," I admitted. "But if you come with me, I'll replace your broken CD players and buy you some new CDs."

"Deal!" cried Skye.

I wanted to settle the birds and get away from my dreams. I dreamed of shredded wings, of bloody bandages, of airplanes trailing clouds of broken songbirds. In the mornings I ran through the woods, trying to shake off the lingering images. I watched for wild birds, but sometimes when I saw one fly by I couldn't tell if it was real or if it was one I had lost.

John agreed to take care of Merlin, the parrots, and the two mourning doves, screech owl, and hairy woodpecker still in the shed. We headed out early in the morning, the kids in the backseat with stacks of books and their music, the birds in comfortable carriers behind them, our overnight bags stacked in the passenger seat beside me. I drove for eight hours, chatting with the kids when they took off their headphones, making a couple of pit stops along the way, trying to think of nothing but the songs on the radio.

The National Aviary is a beautiful and impressive place, with huge open aviaries filled with flowering trees and ponds, and a grand assortment of dazzling birds who willingly approached us when they saw we had seeds or mealworms. On the way out of Pittsburgh we passed several scary-looking motels, settling on a Days Inn on a busy strip that boasted of free cable TV. Driving through the Pennsylvania countryside the following day we saw a sign for ReptileLand and immediately turned off the highway.

"Wow," said Skye. "Maybe we should start rehabbing Galápagos tortoises!"

"They're awesome," I agreed.

"I like those vipers," said Mac. "Would it be a problem to rehab birds and snakes?"

On the way home I thought of all the rehabbers I had met at conferences and on the Internet, many of whom had worked with wildlife for twenty or thirty years.

I had been a home-based wildlife rehabilitator for five years.

Chapter 38

HOPE

The house rang with Jimi Hendrix solos and Eric Clapton riffs, with peals of laughter at midnight from girls' sleepovers. For two weeks the bird calls stopped and John and I hired someone to stay with the kids, went into New York City, had dinner, saw a play, and spent the night, reveling in the strange feeling of twenty-four hours with no responsibilities. Mac found a singer for his band and geared up for another school talent show, dropping a casual announcement the week before.

"I want to get a haircut," he said, causing our jaws to drop.

"No!" cried Skye. "You won't be Mac without your hair!"

"Seriously?" I asked. "You mean now that other kids in school are growing theirs, you're going to get yours cut?"

"Yup." He grinned.

"Cool, dude," said John.

The band played Ozzy Osbourne's "Crazy Train" and just before Mac's solo, his strap slipped off its knob. He reached down and reattached it, beginning his solo late but coming in on the right note and performing the remainder flawlessly. Once again, the ovation fell on deaf ears.

"Mac," I said later that night. "You heard what Ruth said! She was a bona fide rock star, and she said she has never seen a kid so young act so professionally. That strap coming off was not your fault. You fixed the strap and kept going! You were awesome and the audience loved you! Listen to them and don't do this to yourself!"

Occasionally I had short conversations with Maggie and Joanne, or we swapped birds, or, way too rarely, had lunch together. I saw Tanya more than anyone else. We tried to make it a working relationship, and she never failed to bring food or supplies when she brought me a bird; but finally our philosophical differences reached the boiling point.

"I have a wild turkey here from Tanya," said Beth, calling from a small sanctuary two hours north. "She has a badly healed fracture and can't fly. We were willing to keep her, but she's miserable. She's in a big flight cage with other turkeys but she just keeps bashing herself against it, trying to get out. I thought if I gave her a little time she'd settle down, but it's not happening. She's cut herself on the head, she won't eat. I called Tanya, but she said if we won't keep her she'll pick her up and find someone who will."

"Oh, no," I said.

"Our vet will euthanize her," said Beth, "but can you deal with Tanya?"

"Yes," I said.

Steeling myself, I called her. "What?" she cried. "How dare you! What gives you the right to play God?"

"The fact that you're doing a lousy job of it!" I shouted. "She's lost her family and she can't stand being in a cage! Can't you see what she's going through?"

"At least she's alive!" snapped Tanya, and hung up.

I looked out into the woods, my refuge sown with bones. This will be my legacy, I thought. Two hundred years from now anthropologists will dig up my land and think some kind of avian serial killer lived here.

◦　◦　◦　◦　◦

The flight was tarped, the shed prepared for winter. On a chilly mid-November morning I had just taken off my coat when the phone rang. It was Teresa Peppard, who lives in my town and has the uncanny ability to spot an animal in distress, no matter where it might be.

"Hi, Suzie," she said. "Listen, I'm standing here looking at a red-tailed hawk hanging from a tree; I think her leg is stuck in a crack. There's no way I can get to her, she has to be twenty-five feet up. You'd need a really long ladder. I have to get to work, but I can give you directions?"

I shifted into rehabber's autopilot, grabbing gloves, a carrier, a net, and ladder and hopping into my car without thinking of the possible implications. The tree was along a dirt road ten minutes from my house, in a heavily wooded area where houses are few and far between. "The road curves around, then it straightens, then you go down a hill," Teresa had said, using what passes for directions around here. "The tree is on the left, next to an entrance to the Appalachian Trail. People have leaned some walking sticks against it. On the other side of the road are a bunch of dead hemlocks."

When I finally spotted the big female redtail, I reacted with a word not found in family publications. She dangled from a split in the tree by one leg, wings hanging limply. In a futile effort to free her leg she beat her wings and pulled herself to an upright position, only to fall again when exhaustion overtook her. There was no way to tell how long she'd been there.

The ladder I'd brought was nowhere near long enough to reach her, and I didn't have a cell phone. It was a weekday morning, so most people were at work. Where could I get a very long ladder, fast? I had only one idea: the fire department.

On the way over I tried to practice sounding reasonable. "*Hello,*" I recited. "*I'm a wild bird rehabilitator, I live right down the road, I'm licensed by the state and the federal governments; there's a situation with a protected species and I really need your help.*" The problem was that I kept imagining the damage that every extra minute was inflicting on that leg, so I ended up screeching into the parking lot and bolting out of the car in no mood for niceties.

"Excuse me!" I called to the man working on one of the engines. "I have an emergency! Can I borrow your fire truck?"

"Sure!" said the man with a grin. "If it were mine I'd give it to you. But I'm just cleaning it."

"Can I use your phone?" I asked.

I made three calls and reached no one, but meanwhile my new compatriot had used the radio and arranged for the fire chief to meet me at a nearby street corner. The chief would then follow me to the hawk's location, where he could determine what kind of equipment was needed to get the bird down. As the two of us approached the site, however, I could see a very long ladder leaning against the tree, steadied by one man as another climbed up carrying rope and a bag. It was Rich Anderson and Eric Lind, who hadn't answered their phone at the local Audubon center because they were en route to rescuing the hawk.

Rich climbed to the top of the ladder and roped himself to the tree. Balancing precariously, he somehow managed to reach up, extricate the hawk, and put her in the bag, all the while avoiding her thrashing talons. I opened the bag to transfer her to a carrier, and was amazed to see that her leg was bloody but not broken.

"There's a knothole with a crack at the bottom," said Rich. "I'll bet she stuck her foot into the knothole after a squirrel and lost her balance."

I put the carrier into my car and waved good-bye to the whole rescue squad—Rich, Eric, and the fire chief, Joseph Surace, who by this time had been joined by a passing member of the state police. I borrowed a cell phone and called Croton Animal Hospital.

"Dr. Popolow is here," said Charlene, "and she says you can bring the hawk right in."

Carol inspected the hawk's leg, sedated her, X-rayed her, and stitched her torn skin. When I returned several hours later, she pulled the towel curtain away from the front of my carrier, revealing a quietly perched redtail who sported a thick blue bandage on one leg. Carol snapped an X-ray into the viewer and turned off the overhead light.

"This wasn't the first time she's gotten into trouble," she said, tapping the X-ray in three places. "Look—she was shot."

The redtail's ghostly skeleton appeared dreamlike, her pale curved organs surrounded by graceful lines of bone, filmy contours of muscle, and cartilage intricate as lace. The leg that had held her captive was intact. But lodged in her thigh were two shotgun pellets, colorless and unyielding, the jarring evidence of another narrow escape. A third jagged piece of metal rested above her eyes.

"Right now the pellets don't seem to be doing her any harm," said Carol. "I'm assuming they're not lead, since this is an old injury and she doesn't seem to be suffering any effects of lead poisoning. The usual protocol is to leave them where they are—especially when you're dealing with the head, you're apt to do more damage digging around to get them out than you would if you left them alone. Think of all the war veterans who walk around with shrapnel in their heads.

"The leg is another matter," she continued. "There were no fractures, and she can stand up and seems to be able to perch. But at this point I can't tell if there was muscle or nerve damage, and I don't know if there was damage to her blood vessels, which would compromise her circulation. As long as she puts up with your handling her, we'll just have to wait and see."

I put her in the large crate, newly cleaned and disinfected. Not surprisingly, she wouldn't eat the day of her ordeal or the day after; since she was otherwise healthy and in good weight I left her alone, except for the occasional quick check.

On the third morning I took a medium-size defrosted rat, held it in front of her, and shook it back and forth to simulate struggling. I put it on the floor of her crate, shut the door, and by the time I was halfway out the door of the shed she had hopped down from her perch and was holding it in her talons. She's a trouper, I thought. She's going to be just fine.

Chapter 39

DUSK

Treating a leg injury in a raptor is easier said than done. Most raptors don't appreciate people messing with them and will show their displeasure by attempting to grab the offender with their powerful feet. Annoyed wild raptors may be handled with minimal risk by wearing a heavy pair of leather gloves; however, thick gloves impede dexterity, which is needed in order to remove a bandage and clean a wound. I have run into complications when working alone, and I need to simultaneously hold the raptor, remove the bandage, and clean the wound. During this process I inevitably forget to lay out a key item, which means I must then hold the raptor, remove the bandage, clean the wound, and somehow lean across the examination table to retrieve the item, where it is always hiding in the very back of my least accessible drawer. Whenever I am treating a large raptor I think longingly of the Hindu goddess Gayatri, who has five heads and ten hands.

Since John was enjoying a stretch of time working at home, he kindly resigned himself to the daily ritual of holding the big redtail while I took care of her leg. One day John, who had a sore back, decided to hold her upside down on his lap instead of on the exam table. In one of my less intelligent moves I agreed, covering her eyes with a small towel and positioning her so I could see her leg. She seemed relaxed and comfortable, so it came as a surprise to both of us when she suddenly gave one mighty struggle, levitated into the air, and landed right side up on his lap. This meant that there was nothing between his leg and her formidable set of talons but a thin layer of ancient blue jean. We both froze. She briefly looked around, then hopped off his lap and onto a blanket on the floor.

Louise Erdrich once said that the Ojibwe use a single word to describe a man who falls off his motorcycle with a pipe in his mouth and the pipe stem goes through the back of his head. I left the shed wondering what the Ojibwe word would be to describe a man who sits in a chair with a hawk in rehab on his lap and the hawk does a flip and all eight talons go through the man's thigh.

The toes below her injured leg were cold, indicating a possible circulation problem. Each day I treated her leg and then gave her foot and upper leg a massage, silently willing her not to levitate out of John's grip and grab my bare hands with her talons. By the third day her toes were warm but becoming stiff, so I kept it up for another week. Each day her foot would start off clenched and end up limp, proving to me that nobody, whatever the species, can resist a good foot massage.

Her leg healed well although I worried about one of her toes, which curled slightly when she perched. She was a classic redtail—big and fierce but matter-of-fact, accepting her daily handling without becoming either aggressive or frightened. Whenever I looked into her crate she'd stare steadily into my eyes. My dark dreams tapered off, and I dreamed of her flying.

Two weeks after her arrival I moved her into my flight cage. Although there was still a slight curl to her toe, she could fly the twenty-foot length and land on a perch with no trace of discomfort. I wanted her to spend several weeks at a facility with a larger flight cage; if her leg held up I would release her after that, as she was an adult with her own territory, an already proven hunter who needed to go home. Paul had space for her, but at the last minute another friend offered to put her in his big new flight cage with his three redtails. When I released her into the flight she took to the air and landed gracefully on one of the highest perches. I left her exhilarated and optimistic, sure that within a few weeks she could return to her home.

On Winter Solstice there was a full moon and we heard barred owls calling in the woods. We built a small campfire in the field next to our house, huddling in our parkas and staring up at the stars while Mac began a story. He continued until he reached a cliffhanging plot point, then passed it off to Skye. Skye improvised and passed it off to John, and he to me. We went around and around, each of our contributions becoming increasingly outlandish, until we all voted to herd the characters onto a spaceship and send them to the tenth dimension.

Just before Christmas the friend who was taking care of the redtail called, saying her leg looked fine and that she was grasping and landing perfectly. He was about to leave on a trip but his daughter was caring for his birds, and I could pick her up any time. A week after our fun and festive Christmas I was back in my friend's flight cage, staring up at the redtail perched high above my head.

"Oh, my God," I said to his daughter, who stood beside me. "What happened to her leg?"

"I don't know," she replied, visibly upset. "Dad said they were all healthy; all I had to do was feed them, and I . . . I didn't notice anything was wrong until now."

I caught the redtail and looked at her leg in disbelief. The skin around the wounded area had sloughed off, leaving what little flesh remained around the bone hard and black. Her foot was swollen and clenched into a club.

I drove straight to Croton Animal Hospital. "It's not good," said Carol Popolow, eyeing the redtail's leg. "It looks like her blood vessels were damaged."

"Did I put her in the flight cage too soon?" I asked. "Should I have kept up the massage for longer?"

"I don't know that it would have helped," she said.

I stared at her mutely, silently pleading with her to give me a shred of hope instead of the clear, professional assessment that I always asked for, that she always delivered. She must have sensed my desperation, as she somehow combined the two.

"If you want to, you can treat it for a week and see if it responds," she said finally. "Even if it does respond, it doesn't mean there won't be other issues. But you can make another decision in a week."

"Uh-oh," said John, after I'd settled the redtail for the night. "You look like you could use a nice big glass of wine."

I had one, then another, and eventually lost count. "Excuse me, whoever you are," I said, gazing blearily at John at the end of the night. "But what was the reason for this fine evening? Was there some sort of bird calumny? Caluminimy?"

"I believe the word you're looking for is 'calamity,'" he replied. "And not to worry, all the little birdies are snug in their nests. The only calamity around here is what's happened to your vocabulary."

The following day, head pounding, I began the redtail's treatment. Each day John would hold her while I carefully washed her leg, dressed it with a special ointment, then padded and wrapped it. Even though we usually referred to her as "the redtail," we christened her CJ, for Calamity Jane.

By the end of the week her leg was responding, the blackened scab eventually falling off to reveal pink tissue underneath. Despite her ordeal her appetite remained constant, allowing me to hide her pain medication in her food and convincing me that she had the will to continue. But then her toes began to curl, signifying that her tendons were contracting, so I splinted two of them to keep them straight. Her moods shifted. Sometimes I would approach her and she'd stare steadily back into my eyes, positive and strong; other times her depression would hit me like a blow. The ability to feel the mood of an animal is a sense that is strong in some people and weak in others, and can be honed by time and experience and desire. "I promise you," I whispered to her, even though I never make promises I don't know that I can keep. "I promise I'll let you go."

She needed more exercise than the large hospital crate could afford her, but she was not yet ready for the flight cage. I began placing her on my heavily padded examination table, upon which rested a thick round section of log.

She'd hop onto the log and I'd roll it slowly back and forth, giving her the opportunity to use her toe and foot muscles to grip and balance. One day she hopped off the log and stood on the table.

"Come back up here," I said, and tapped the log with my hand. After a moment, she jumped back onto the log. Slowly I reached down to the floor and picked up a wooden perch, which I placed at the other end of the table. When I tapped it, she gathered herself and jumped from the log to the perch. When I tapped the table, she jumped down. Graced by her acceptance of me I'd watch silently as she rocked back and forth on her perch, her great curved talons inches away from my unprotected hand.

Then one day she jumped onto the log and quickly turned around, ripping off her entire back talon and staining the white towels red with blood. I washed and padded and wrapped and then wrote a frantic e-mail to Louise Shimmel, who runs the Cascades Raptor Center in Oregon, and who quickly sent back several paragraphs of information about talon regrowth. "You'd be amazed at how often that happens," Louise concluded. "Carry on and don't beat yourself up over it."

But of course I did, berating myself for not having a mid-size enclosure that would allow a recovering bird to move about more freely than the hospital crate. My dark dreams returned. Later I called Eileen Wicker, who runs the Kentucky Raptor Center and is another of the Grand Masters of the Raptorcare list, desperate for any piece of advice that might make a difference. "Take down this number," she said. "They make a healing cream that's the best thing I've ever used. But talons take a long time to grow back," she finished. "You know she'll be with you for at least a year."

I alternated between letting her exercise on the table and putting a second padded perch in her crate. When she was on the table she would occasionally bang her bandaged talon sheath and make it bleed, but if she stayed in her crate her tendons would begin to contract and her toes to curl. I massaged her leg and toes, trying to believe that I could heal her damaged limb through sheer force of will. Increasingly restless, she eyed the windows and made me fear that she would suddenly launch herself through one of them, to try to escape a life over which she had no control. At the beginning of the third week I called the Raptor Trust and talked to Kristi Ward, whom I had met once and spoken to many times over the phone. I relayed everything Dr. Popolow had said, describing the situation calmly until the end, when my voice began to betray me.

"Kristi," I said. "I would do anything for her. But I'm so tired. I'm afraid I'm not making the right decisions. I want what's best for her. I don't know if I'm doing the right thing."

"Of course, we'll take her," said Kristi. "Bring her down."

That afternoon CJ slipped off the table, dropped to the floor, and broke off another talon.

I took her to the Raptor Trust the following morning and drove home with an empty carrier, searching for something more I could have done for her, wondering if I should have done as much as I did. I wrote Ed a long stream of consciousness, ending with an anguished mea culpa. "Dear Suzie," came the immediate reply, "Shakespeare said in *Measure for Measure*: 'Our doubts are traitors, / And make us lose the good we oft might win, / By fearing to attempt.'" When I finally slept I dreamed of a forest where redtails hung from the trees, struggling to set themselves free.

Several days later I called the Raptor Trust. "We put her in the clinic," said Kristi, "but she was really nervous and upset. She started bashing and knocked off another talon, so we put her outside in a small flight."

I'm going to lose her, I thought.

"While she was in the flight she knocked off her last talon," she said. "Our vet examined her and said her circulation had been too badly compromised. She's not going to be releasable. I'm really sorry."

"Is she in pain?" I managed.

"She's not putting any weight on that leg," said Kristi.

"What do you think is the best thing for her?" I asked, already knowing the answer.

"Our vet advised us to euthanize her," she said quietly, "but we'll do anything you want us to do."

I promised I'd let her go.

After I hung up the phone I put on my running shoes, called Merlin, and ran to the top of a ridge high above the Hudson River, high above a world for the most part oblivious to the death of a thousand redtails, let alone one. I sat on the frozen ground and hugged my knees, my body knotted and aching.

Bring them back, then let them go.

I brought her back. Twice. And for what?

I returned red eyed to the house, where John and the kids provided sympathy and consolation and sent me to bed early. Exhausted and defeated, I slept a black and dreamless sleep for ten hours and woke up in tears. Pulling myself together, I awakened the kids and readied them for school and walked them down to the bus, then fell apart when I returned to the house.

For the next few days I held myself together around my family, none of whom were fooled. I assured them I was all right while trying to stave off a growing sense of panic. I had no wild birds to care for, but kept thinking there was one I had forgotten. I stared straight ahead as I ran through the woods, expecting to see CJ flying beside me out of the corner of my eye, but she wasn't there. I spotted a redtail soaring above the field, but it wasn't her. I went to sleep hoping for a dream, even if it was a bad one, but my dreams had vanished.

All four of us sat in front of the fireplace one night, watching a movie in which one of the characters moves to Montana. The sky was blue and endless and empty. I searched the screen for a red-tailed hawk, my heart pounding.

She wasn't there, either.

I rose abruptly and hurried through my bedroom and into the bathroom. I huddled on the floor in the corner, crushed beneath the weight of a world filled with wronged and wounded creatures, a world I stupidly thought I could make right just because I cared so much.

I heard John's footsteps, and the door opened slowly. "What is the matter with you?" he said.

"She's gone," I sobbed.

"Who?" he asked. "Who are you talking about?"

"The redtail," I said. "I've looked everywhere and I can't find her."

John's expression darkened. "But . . . she's dead."

"Why can't you understand?" I cried.

Chapter 40

SHADES OF GRAY

"You're a fanatic," said Elisha Fisch, Ph.D.

"Am not," I said.

Dr. Fisch leaned back in his comfortable chair, glancing briefly at the note-book resting on his lap.

"Think about it," he said. "You believe in something so totally that you have allowed it to take over your life. It causes you distress, yet you don't stop or even cut down. You have isolated yourself to the extent where you rarely see your friends. You believe your way is the right way, and that anyone who disagrees with your philosophy is wrong."

"Hmmm," I said.

"Tanya. Can you see her point of view?"

"No."

"Why not?"

"Because I disagree with her and she's wrong," I said, only half facetiously.

"Take a minute and try to understand why she would have kept your hawk alive."

I blinked rapidly. "Probably for the same reason I did. Because she couldn't stand the thought of losing her."

"But eventually you had the hawk put to sleep."

"And here I am."

"Well. It wasn't exactly an isolated case of cause and effect. But I suspect that Tanya might say that if, at the end of the day she feels good and you're in pieces, you are misguided. And self-destructive."

"And I would say if you're not doing what's best for the bird, you shouldn't be in this business."

"Have you told your friends on the mailing list about what happened with the hawk?"

"No."

"Why not?"

"I don't know."

"You said everyone on the list consoles one another when something goes wrong."

"They do."

"Have you told anyone about your hawk?"

"No."

"Why not?"

"I don't know. Maybe I'm becoming birdlike. I don't want to show any weakness."

"I thought birds don't want to show weakness because they might be spotted by a predator, who would then try to eat them."

"It's not an exact analogy."

"It might be closer than you think. Do you consider yourself weak?"

"Do I seem like a tower of strength to you?"

"Actually, yes."

"Then why can't I tell people no? Why can't I turn birds away? Why can't I take care of them and just soldier on when they die? How do I stop caring about them?"

Dr. Fisch regarded me with such compassion that I had to look away.

"There are several things you can do to cut down on the number of birds you take in," he said. "Right now you might consider shutting down for a few months. How do you stop caring? If you ever figure that out, let me know."

Later I drove home and changed the message on my answering machine. "If this a wild bird call I'm sorry, but we are temporarily closed," said my recorded voice. "Please call the Wildlife Hotline." I called the animal hospitals. "Just take it easy for a while," said Carol Popolow. "We all know how hard you've been working." I left both my electronic mailing lists, and my rainbow-colored library of three-ring binders lay idle.

The February woods were gray and silent. I haunted the ridgelines, running the border between earth and sky, trying to achieve a clear mindfulness. I sat and watched the river, stroking Merlin's big, powerful shoulders, knowing his strength was an illusion, knowing that no matter how healthy my family seemed they were only a stroke of luck away from disaster. I was no longer exhilarated by the beauty of a soaring hawk; as tears stung my eyes, I felt defeated by its fragility and retreated back into the woods, where the winter canopy obscured the sky.

I watched Mac cradle his guitar as he played "Hey Joe," saw Skye place her camera carefully on the shelf when she finished using it. At least a guitar or a camera won't sicken and die on them, I thought, grateful they could spend their love on something that was, for now, unable to hurt them.

Skye took a series of photos of the snow-covered field shrouded in mist, dreamlike images in shades of gray and brown. As I stared at them I tried to envision myself being drawn into the photograph, hoped the outline would seep into my subconscious enough to return that night as a dream. But nothing appeared while I slept. I listened to Mac play the opening notes to "Stairway to Heaven," remembering when teenaged boys would try to decipher the meaning of life from its lyrics. When I was seventeen I didn't need Led Zeppelin to explain the meaning of life to me, but now I wasn't so sure.

Ed sent me e-mails. "Call me if you need to talk," he wrote. "You will get past this point in your life, it will just take a little while."

I willed my rickety psyche to a feeling of calm normalcy. I stopped answering the telephone, and turned the sound off on the machine so I couldn't hear who was leaving a message. John sorted through them all at the end of the day, relaying the ones from friends and family and deleting the rest. Inevitably, though, someone left the sound on by mistake. I was reading in the living room and heard a woman's voice coming from the kitchen.

"Please help me," it said. "I have this beautiful bird here, and he's really badly hurt. I don't know what to do. Please call me back."

I hurried out the back door and fled to the rocky outcropping behind the house. I looked down at the shed and the flight cage, on the small world I had so carefully created.

It wasn't supposed to end like this, I thought.

The following morning I changed the message on my answering machine. "If this is a wild bird call I'm sorry, but Flyaway is permanently closed," said my recorded voice. "We will not be opening again."

◦　◦　◦　◦　◦

I zealously vacuumed the house, cleaned out all the closets, rearranged the kitchen and garage, caught up on five years' worth of yard work, and trailed the kids until they begged me to stop. I filled the bird feeders, then tried to avoid looking out the window lest I start to wonder about the songbirds I didn't see. I ran through the woods with Merlin, trying not to think about anything.

"I wish I could see what's going on in that head of yours," said John.

"No, you don't," I said, managing a small smile. "It's like Iwo Jima in here."

As spring eased its way toward us I watched as Nacho strutted across the yard to investigate an abandoned bone of Merlin's, wondering how many of the migrant birds would return this year and how many would fall by the wayside.

"Mom," said Mac, startling me. "Don't do this to yourself."

"What?" I said.

"You're such a good rehabber," he said. "You do such a good job. When things go wrong, it's not your fault."

I met his eyes.

"Thank you," I said.

⊙   ⊙   ⊙   ⊙   ⊙

My friend India Howell left her orphanage in Tanzania for the first of her biannual fund-raising trips to the United States. As always she rolled up the driveway in her rented car, here for a twenty-four-hour visit, and from deep within
her enormous suitcase came handmade gifts: placemats, baskets, a kikoi for
her goddaughter Skye. She opened her laptop, the African Children's Chorus
burst into song, and photos flashed on the screen: laughing kids of all ages,
their arms around each other, around volunteers, gathered in front of the new
library, the new clinic. Kids who suddenly, magically, had a future.

"I get away," she said, as the two of us drank white wine at a small restaurant overlooking the Hudson River. "Twice a year I come back here for four
weeks at a time. I have a whole network of people helping me—not that they
don't come with their own complications."

"But you love those kids. How do you say no?"

"I don't have a choice. Every single day I have people lined up at my office
door holding kids. They tell me they can't feed them, they have no money, and
if I don't take them the kids will die. But my resources are limited. And if I take
in more than I can care for, the whole organization will come crashing down."

"Yeah," I said.

"You know that saying—'You can't save them all'?"

As she regarded me, her expression changed. "You never really believed it,
did you?" she asked.

"Damn," I said lightly, trying to force a smile. "The problem is I've always
seen the world in black and white."

India grinned and raised her eyebrows, as she did whenever someone stated
the obvious. "Evidently it's not working for you anymore," she said. "You think
it might be time to let in some shades of gray?"

◎ ◎ ◎ ◎ ◎

A few weeks later I heard Wendy's voice on the answering machine. "I know you're not taking birds anymore," she said, "but I just wanted to give you the option. I have an orphaned crow."

I stared at the machine. "You don't have to give me an answer now," she continued. "Just think about it. I'll keep him for another day, and if I don't hear from you by tomorrow night I'll give him to a rehabber up north. There's nothing wrong with him—he just needs a place to grow up."

I considered the idea from my small corner in limbo, trying to decide if a nestling crow would be a life raft or a packet of tainted heroin. You have no emotional discipline, I told myself. Maybe if you could just exhibit a little clinical detachment, everyone around you would be better off.

By the time I picked up the little crow he had decided that humans were hospitable creatures, good for warmth and food and the occasional head scratch, and had made himself right at home. As it turned out, Wendy had found him herself. Walking her land after a violent thunderstorm, she spotted a nest of crows on the ground. Two were dead; the third looked up at her through wide blue eyes, and moments later found himself cradled inside her sweatshirt. She waited. She left and returned to the spot later, but there were no signs of any adults.

I thought perhaps he could be returned to his parents if we rigged up a makeshift nest, but a steady rain had developed and the chance of a reunion seemed slim. "I won't call you for any other birds," said Wendy. "I just wanted to give you the choice with this one."

I brought him home to a chorus of approval. The kids peered into the box, exclaiming over the rare white secondary feather on the crow's left wing.

"Maybe this will be a good thing," said John.

I sat down with the phone and in three calls found a companion: just that morning a rehabber west of me had discovered a box containing a young crow on her doorstep. The attached note explained that the crow had appeared mysteriously in a driveway, with no other crows in sight. There was no return address, which meant no chance of reuniting the fledgling with her family.

While Wendy's crow was a round, easygoing little preteen, the new one was a tall, thin, and fearful adolescent. I set them up in a roomy enclosure containing a nest basket, various sizes of branches for perching, and a potted plant for the fearful one to hide behind. "What are you going to name them?" asked Skye.

"I'm not giving them names," I said.

"Then how are you going to tell stories about them?" asked Mac.

"I'm not," I said. "I'm just going to feed them and leave them alone. If I have to refer to them, I'll call them One and Two."

"One and Two!" cried Skye. "Are you crazy?"

Immediately both kids grimaced, clenching their teeth and bugging their eyes in dismay. "She didn't really mean you were crazy," said Mac. "Maybe a little upset, but not crazy."

"That's right," said Skye. "Mom is definitely not crazy." Turning toward Mac, she whispered, "Unless she names them One and Two."

Soon after I approached the crows' enclosure holding a bowl of food. I moved slowly and quietly, eyes downcast, intent on showing the new one that I posed no threat, but it didn't work. She let out a gasp of fear, dove behind the potted plant, and stood quaking while the nestling did everything to mime starvation but suck in his cheeks and point to his stomach. He staggered to his feet, opened his beak like an alligator, and let out a barrage of hoarse, wheezy begging cries, instantly triggering my innate feed-the-crow response. Whipping out an eight-inch pair of tweezers, I stuffed bite after bite into his open beak until he turned away. I offered a bite to the fearful one, but she would have none of it.

Maybe One will be a good influence on Two, I thought, immediately feeling ridiculous and deciding that my kids were far more sensible than I. Hoping the little crow pair would develop a camaraderie like that of the old comedy team, I christened them George and Gracie.

"But this doesn't mean I'm going to fall for you," I said to George, who gazed at me with admiration. "So don't get any ideas."

An hour later I tried again. As before, George was thrilled and Gracie was aghast. Gracie was already thin and couldn't afford to lose more weight, but the last thing I wanted to do was force-feed an already terrified young bird.

"She's not buying it," I told John and the kids dolefully. "I might as well have ridden in on a motorcycle with a bone through my nose."

"A *bone* through your *nose*?" Skye screeched with delight. "How the *heck* couldja get a *bone* through your *nose*?"

"Gracie's not against you personally," said Mac soothingly. "She just doesn't like what you represent."

"Don't worry," said John. "She'll come around."

A half hour later I repeated the process with the same result, although Gracie was a bit less frantic. Another half hour later I gave it one last try, and to my relief Gracie lunged through the plant foliage for the tweezers, snatched a piece of food, then retreated behind the greenery. Hunger and George's good example had won her over.

During the next few days Gracie unwillingly settled in, and George left his nest and began hopping around the enclosure. Their initial views of captivity never changed: George openly appreciated my hospitality, while Gracie made it clear that she was a prisoner of war and was simply biding her time until she could stage a daring escape. A week after they arrived I began putting them in the Crow Mahal during the day so they could enjoy the sun and the outdoors. I filled it with crow toys, several perches, a plate of food, and a large dish for bathing. I sat in the hammock nearby, watching.

I thought of Lena, the first crow I had ever known, for whom I could still feel sharp pangs of loss even though years had gone by. I thought of the songbirds, the herons, the redtails, all the ones who had followed her despite my best efforts. For days I opened the door to the Crow Mahal, my head crowded with lost birds, only to have George effortlessly nudge them aside. I was determined that he bond with Gracie, and not with me; but evidently George believed there was room for both of us, and his will was far stronger than mine.

One afternoon I watched them from the deck, so engrossed that I startled

at the ring of the telephone. Once again, someone had forgotten to turn the sound off. "I heard you had closed down, but I have an injured blue jay and I wanted to see for myself," came Tanya's voice on the answering machine. "I should have known! You always acted as if you cared about these birds, but when you come right down to it you won't lift a finger to help them!"

A small knot twisted in my stomach. I heard the sound of splashing and lifted my eyes to the Crow Mahal, where George and Gracie were vying for space, each energetically dipping their heads, shrugging their wings, and wagging their tails as water droplets flew in all directions. Slowly the knot began to loosen.

"I guess you can't please everyone," I murmured.

Chapter 41

BREAK OF DAY

"Look at that crow," said Mac, putting his cereal spoon down and pointing through the kitchen window. "He's limping!"

Five crows had landed on the slope and were busily pecking at the seeds I had scattered. I could identify Nacho by his voice and behavior, but I couldn't tell Lo and Behold from the wild crows who sometimes came by. One of the crows was definitely limping; all I could say was it wasn't Nacho. I grabbed my binoculars, but couldn't see any sign of a wound.

"Maybe something bit him," said Skye.

The slope was a busy place that morning. Songbirds flew to and from the feeders; in addition to the crows a pair of blue jays, a squirrel, and several chipmunks were combing the soil, looking for food. As the four of us watched, the crow limped close to one of the chipmunks, studiously ignoring him, seemingly engrossed in his own hunt for seeds. But suddenly the crow jerked to the side, grabbed the chipmunk, and took off, the chipmunk struggling mightily in his beak. We all watched, wide-eyed, as the two of them disappeared over the hill.

"What does he think he is, a hawk?" I exclaimed.

"The poor chipmunk!" cried Skye. "Is he going to kill him?"

"Have you ever seen a chipmunk's teeth?" said Mac. "Now that crow will be limping with both legs."

In less than a minute the crow returned, chipmunk-less, and resumed looking for seeds.

"When you release George and Gracie, tell them to stay away from that bird," said John. "He's a bad influence."

George and Gracie were almost eating on their own, and they spent most of their day outside in the Crow Mahal. One day I arrived with a large rectangular child's mirror. George strode up to his image, gave it a few hard pokes, peered behind it, and finally pushed the whole thing over onto Gracie, who slithered out from beneath it and took refuge on one of the perches. After I righted the mirror and tied it to the side of the enclosure, Gracie approached it suspiciously. For a few moments she stood before it, immobile; then she walked over to her food dish, picked up a grape in her beak, returned to the mirror, and studied herself from different angles. She dropped the grape, picked up a strawberry, returned to the mirror, and repeated the process. From then on whenever I introduced a new food item Gracie would pick it up, carry it over to the mirror, and observe herself holding it before she ate it. George quickly caught on, although he was never as methodical as Gracie. Once George impaled an apple slice with his lower beak and couldn't get it off. He walked over and peered into the mirror, wobbling his apple-festooned beak up and down until my laughter destroyed the mood and he slapped it off with his foot.

⊙ ⊙ ⊙ ⊙ ⊙

By the time both crows were self-feeding their eyes were slowly starting to change from their juvenile blue to adult brown. It was time to move them into the flight cage, where they would have space to explore and begin to fly. Both crows seemed delighted by their new surroundings: George had more room to play, Gracie more room to avoid human contact. I wondered if George would begin to distance himself from me, as most of the other formerly friendly young crows had done once they entered the flight cage, but he stuck to his guns. Whenever I walked toward the flight cage I'd see the two crows in the middle of some companionable interaction, but as soon as I entered George would fly toward me and Gracie away. George would land on a nearby branch and peer into my face, pull at my clothing, and offer me crow toys. Had I done things by the book I would have ignored him, cleaned up, replaced food and water, and left—but I could no more ignore George than I could fly to the moon. I'd say "Hello, George!" and George would snort and reply *"Ahh-low! Ahh-low!"* I'd mimic him, he'd mimic me back, and we'd carry on like a pair of happy drunks while Gracie watched disgustedly from her high perch.

The bird calls tapered off and I turned the volume back up on the answering machine, although I still never answered the phone myself. One afternoon as I was listening to the messages I heard Jayne's voice.

"I know you never answer your phone anymore," she said, "but have you ever raised a damned Cooper's hawk? There's a nest of them behind my house and something must have happened to the parents because they've been screaming their heads off all day. I know I should just ignore them . . ."

I listened in amazement as she finished helplessly.

". . . but I can't! They're hungry."

My surprise turned to recognition. I remembered drawing my own lines in the sand, permanent and unyielding, until unexpected gusts of wind blew them all away.

"You're probably wandering around the woods somewhere, so never mind," she finished. "Hope you're doing okay. Call me later."

I released George and Gracie one sunny morning. I had done all I could to get them ready: they recognized local wild berries, they could chase down crickets, and they dug enthusiastically into various dead wildlife I scavenged from other rehabbers. George was fine with me, but along with Gracie he would panic if a strange person entered his flight cage. I had no reason to keep them, save for the quick rush of despair I felt whenever I thought about losing the bird I had sworn not to fall for.

When I released them, Gracie, true to herself, took off like a bullet. George, close behind her, disappeared but eventually returned alone. He spent the day flying around the house, landing on the roof, the deck, various branches, and occasionally, my head. "Get off! No bird hats!" I shouted, even though rules mean nothing to a crow. From time to time I'd spot Gracie's slim silhouette soaring over the house, once in the company of a crow I thought was Nacho, but she never landed where I could see her. At twilight George settled in the black birch outside the kitchen window, where I kept returning to check on him until darkness hid him from view. The following morning I rose early and walked around the house with a plateful of crow food, peering up into the trees, willing myself not to worry; but when George finally coasted down out of the sky and landed on my shoulder, I was dizzy with relief. He mumbled a few soothing crow phrases, leaned against the side of my head, and fell asleep.

Like an avian anthropologist, George spent the summer flying back and forth among Gracie, the other crows, and his adopted human tribe. Gracie remained a shadowy figure, occasionally spotted grabbing her daily meal from the feeding platform or perched on a far-off limb with George, but she never came near the house. And even though George quickly accepted John, Mac, and Skye, he would have nothing to do with friends or visitors. At the sight of a friend's dog he would frantically circle the house, screaming the crow danger cry, then vanish into the woods; but after a few days of studying our interactions, welcomed Merlin as a valuable playmate. George would sit, hidden and unmoving, in the shadows of the spruce tree and watch as the unsuspecting Merlin meandered across the lawn. Suddenly George would launch himself, roar up behind Merlin, and drag his feet across the top of his head, sending Merlin diving for cover until the poor dog realized that once again he'd been had. Flying low, George would careen around the house with Merlin in hot pursuit, forcing any innocent bystanders to hurl themselves out of the way. Merlin seemed to be no match for the devious George until he discovered George's secret passion: anting.

Once George discovered anting—although, as with the blue jays, I'm not exactly sure how he discovered it—he was hooked. He'd poke around through the weeds until he found an anthill, then he'd roll on top of it and gyrate until he was covered with racing ants. Once in a while George would have such a good time that he'd lose track of his surroundings, giving Merlin the chance to thunder up like a draft horse and scare the wits out of him. George, aggrieved, would retreat to the spruce until he recovered his aplomb. What goes around comes around, I wanted to tell him. Quid pro crow.

No one could bring out George's exuberant side like Mac, who would race outside, throw himself down on the grass, and wait for George to land beside him. Mac would pick up a twig and brandish it at George; George would open his beak and let out a guttural *caw*, then flop onto his back and bicycle his legs while Mac grabbed his feet, scratched his belly, and tossed him dandelions. Eventually George would jump up, snatch the twig, and fly to the trampoline, where the festivities would resume. While Mac jumped George would hop from pole to pole, commenting loudly, or grab a hanging strap by his feet and swing back and forth upside down.

His relationship with Skye was more complicated. When Skye climbed into the hammock with a book George would settle next to her, drowsy and content, until he decided it was time to get up, stretch, and poke Skye in the ribs or try to rip the book's cover off. But as much as he liked relaxing with her, he like playing practical jokes on her even more. George's gold standard was when Skye and a friend were playing in the empty parrot flight enclosure next to the house. Even though the door was closed it was slightly off-kilter, so in order to lock it from the outside you'd have to push it another inch to a fully closed position, then slide the dead bolt shut. George must have watched me lock the parrots inside dozens of times and seized the opportunity: landing in front of the cage he climbed up the wire front, beating his wings in order to push it the final inch to a fully closed position, then he slid the dead bolt home and successfully locked the girls inside. John and I arrived moments later, drawn by their shouts for help.

"Who locked you in?" I asked, puzzled.

"It was George!" they chorused.

"There's no way he locked you in," said John with practiced skepticism. "No way."

"Go ahead, Daddy," said Skye, filing out with her friend. "He'll do it to you, too."

John entered the flight and pulled the door shut behind him while we all watched from the deck. Within a minute George had repeated the process and John was shouting for help.

"Don't look at us," I called to John. "Ask George to let you out."

The only one unimpressed with George was Mario, who dislikes other birds in general and crows in particular. Mario spent a good part of the summer glaring evilly at me and shouting "War!" at George. As it turned out, he was also plotting.

For a brief period Merlin figured out how to open the screen door to the kitchen. He'd push his nose against the handle, slide it open it just wide enough for his body, and then slink into the house, leaving the open screen as evidence. Had he closed it behind him I would have let him get away with it, instead of eventually removing the handle. But during that small window of time Mario began calling to George in my voice. He'd call *"George! Geooooooooorge!"* enticingly, and a few times I walked into the kitchen from the living room just as George was about to walk in from the deck. I'd say "Get out of here, George, you're not a house crow," and shut the screen, never suspecting that I was thwarting Mario's plans. But one day I entered the room just in time to see Mario standing on the kitchen table while George walked through the open screen door. As George strode by, Mario pushed a heavy plate off the table, missing George by inches and sending him flapping back out the door. Mario regarded me as I swept the shattered plate into a dustpan.

"War," he said.

We drove to Randall's Island in Manhattan to see Cirque du Soleil, something we could never have done in summers past. I watched as the dazzlingly costumed performers defied gravity and thought: if they can achieve perfect balance, why can't I?

The following morning I whistled for Merlin, crossed the field, and headed into the woods for a run. I had taken only a few steps when I heard a "whoooshhh" and felt a gust of air over my head. I ducked down, thinking it was the female goshawk, even though I knew she had abandoned her troublesome old nest and moved deeper into the woods. It was George.

"Do you want to come with us, George?" I called, and as I ran down the trail he caught up and flew beside me, and I ran faster and faster until once again I felt as if I were flying. That night I sank into my black sleep for only an instant before the images returned in a tumultuous blaze of form and color.

Chapter 42

SONGS OF REDEMPTION

George became our hiking companion. Mac and Skye and I would tramp off through the woods, with Merlin galloping before us and George strafing us from behind. Instead of sauntering slowly through Blue Jay Town, looking for Harry and his jaunty descendants, we'd race through and try to avoid the screaming blue demons who would mob a single crow. The lone turkey vulture sunning himself at the top of a tree no longer enjoyed our silent observation; we'd give George noisy encouragement as he rocketed into the air, circling and cawing loudly at the intruder. We patrolled the edge of the pond, combed the forest floor for bugs, felt pangs of anxiety at the sight of a red-tailed hawk, and viewed the world from the eyes of a young and adventurous crow.

Near the end of August we were scheduled to go on a week's vacation.

"I can't go," I said to John. "I can't leave George."

"Balance," said John. "A vacation will help you achieve the balance you're always talking about. George will be fine with Bill."

Bill Whipp is a cheerful, world-traveled Wisconsin native who settled in the Hudson Valley and spends some of his retirement time house- and pet-sitting. He loves animals, both domestic and wild, and they love him back. He was intrigued by the idea of continuing to provide a base camp for our varied group of crows.

"Come over a few days before we leave and I'll try to introduce you to

George," I said to Bill. "But I don't know what he'll do. He might not even come near you."

"No problem," said Bill, with his usual sunny optimism. "We'll get along eventually."

And, of course, they did. Even though George would have nothing to do with strangers, once he was introduced to Bill he treated him like an old friend. Every few days I'd check in with Bill from our vacation spot and he'd regale me with tales of George and sightings of Gracie and Nacho, with assurances that dog and parrots were fine, and that everything was going according to plan.

"Mom, I had a dream," said Skye, as we were sitting on the beach.

"Tell me about it," I said.

"You know the Japanese maple in the front yard?" she asked. "I dreamed it was fall and the sun was shining on it and it was *so red*. And all over the tree were blue butterflies. They were electric blue and so bright it was like each one had a light inside it. There was the red tree and the electric blue butterflies, and they were so beautiful, and I thought to myself if I could just photograph it, it would be amazing."

"Sweetie," I said. "It's already amazing."

When we returned George was nowhere to be found. "He was here an hour ago," said Bill. "Then I saw him with another crow, I think it was Gracie, and they took off toward the field."

I feigned nonchalance as I stood on the deck and searched the trees for a big glossy crow with a single white feather. I circled the house, calling his name, wondering if he and Gracie had headed for parts unknown. I walked up and down the driveway several times, then finally turned to see a large black bird flying straight toward me. "George!" I shouted, watching in amazement as he shot straight up into the sky and spiraled down like a corkscrew, spreading his wings a few feet from the ground and roller-coastering back upward. In return I leaped into the air and waved my arms like a madwoman, whooping and carrying on and wishing I were a gymnast so I could express my joy with the earthbound equivalent of George's display. Finally he coasted down and landed on my shoulder, rumbling in his gravelly voice, and my vacation was complete.

I went out and bought a half a dozen new CDs, and once again the house hummed with music. "I guess I didn't notice it until now," said John. "Mac has been the only one playing music around here lately."

Toward the end of the summer George was spending longer periods away from us. He no longer slept in the birch tree by the kitchen window, but would vanish before dusk and reappear sometime in the morning. Although we spotted him with Gracie and with other crows more often, he continued to both entertain and bedevil us. After we installed an electric fence for Merlin, John circled the house, carefully placing the little white flags that marked the fence boundaries into the ground. John had no idea that George was right behind him, just as carefully pulling out each flag and tossing it away.

When the kids started school I settled into a fall routine. I took morning runs with George and Merlin, then sat on a log next to the pond's edge and watched as they explored. I began to consider rehabbing again. I wondered if I could make Flyaway, Inc. into a manageable operation, or if I were shortchanging my family by rehabbing at all; if it were possible for me to grow a thicker skin, or if I would always be an accident waiting to happen.

"People think I'm really tough," I said to George, as he yanked at my shoelaces. "But as you may have noticed, I'm actually kind of unstable."

The kids were getting older and I was spending more time in the car, ferrying them to and from playdates, sports events, and band practices. They needed me less in some ways, but more in others. Sometime in the not-too-distant future, my own fledglings would leave for college.

"If you're working alone, you have to specialize," Wendi Schendel, caretaker of wood and sea ducks, had told me. "You can't be all things to all birds."

Perhaps I could just take in raptors, as I had once planned. But what about the songbirds? What about the herons?

George rocketed into the air and circled the pond, landed again, picked up a stone, and dropped it into the water. I reached down, picked up a stone, and tossed it in myself. How many times had I done that over the last five years?

I returned home, carried a notebook and pen out to the deck, and started making a list. George sat at the top of an old hemlock, preening his feathers. "No matter what," I called to him, "I'll never turn down a crow."

Late one fine September afternoon I found George perched on a birch limb about six feet off the ground. I sat down on the grass, waiting for him to land beside me. This time, though, was different; he stayed where he was, regarding me quietly. Several minutes drifted by, until I realized that he was about to leave for good and was saying good-bye. I whispered a silent plea.

Don't go.

I don't claim to be able to read the mind of a crow. I simply felt a terrible sense of impending loss, and in a moment of panic silently vowed to do anything if this one bird, out of the legions I had taken in, would give up his freedom and stay with me.

But I hadn't raised him to be a captive bird, and I couldn't ask him to be one. I sat still, trying to hide the turmoil I was feeling and let him go. Finally I rose. "I'll miss you, George," I whispered.

"What's the matter?" said John, when I walked into the house.

I returned an hour later and George was gone. I never saw him again.

EPILOGUE

In my dream I hear a rush of wings.

Alone on a hill I am the center of a living kaleidoscope, for wherever I look there are birds in flight. Egrets in majestic slow motion, hummingbirds like the sparks from the tail of a comet. Swifts and kites, owls and sandpipers, green-winged teals and lazuli buntings. The circle breaks apart and flows past me like a shimmering wave, then returns and pulls me upward. And in that breathtaking moment I am no longer an observer, I am a part of them.

Roger Tory Peterson said, "First birds captivate you, then they enslave you." All bird rehabilitators must dream of flight, of being accepted by the creatures who have enslaved us but who normally, wisely, would have nothing to do with us. Our reality is that we are rarely accepted; if we are given temporary grace, it is by a weak and wounded bird who will probably return to fearing us once it has recovered. But it is during those small, elusive moments when a frightened bird no longer panics at the sight of us that we sing in our chains, that we come close to our dreams of acceptance as we try to help it find its way back into the sky.

Although I have written only about birds, every rehabilitator—whether he or she cares for birds, mammals, or reptiles—has stories that would entertain and astonish an audience. All wild animals are complicated, remarkable creatures, and the odds against their survival are growing. They need help.

Wildlife rehabilitators are among the most generous, compassionate, hard-working people I know. Knowing how few of us we are, and driven by the odds stacked against the creatures we care for, we tend to work until we drop or burn out. Please: find the rehabilitator or sanctuary nearest to you, or contact any of the rehabbers mentioned in this book, and write a check or offer your help. By law, those who work with wildlife cannot charge for their services, so everything—food, medicine, equipment, facilities, and the like—must be paid for out of the rehabber's pocket, or by donation or grant. Money, labor, blankets, crates, lumber, electronic equipment, transportation of injured wildlife— the list of needs is always endless. To find your local rehabilitator, contact your State Department of Environmental Conservation or Protection and ask for a list of rehabilitators or sanctuaries, or access the Web site of the National Wildlife Rehabilitators Association (www.nwrawildlife.org) or International Wildlife Rehabilitation Council (www.iwrc-online.org).

No matter where you live, you can send a donation to Ward Stone, New York State Wildlife Pathologist, in Albany, New York. Ward is a lifelong champion of all wildlife. He gathers and catalogues the evidence against environmental poisons, and stands between the human community and any emerging zoonotic diseases, yet he never has enough money to do his work.

And finally, wildlife cannot exist if there is no place for them to live. The biggest threat to wildlife all over the world is habitat destruction. Support your local Land Trust, Open Space organization, or Audubon Society; support the strengthening of wetlands and steep-slopes protection laws; and the next time a local wild area is threatened by yet another greedy developer, join the battle against him.

◎ ◎ ◎ ◎ ◎

They say you should take a mental snapshot and keep it tucked inside your mind, an image you can visualize when things go wrong and you need support. Despite the times I have felt alienated, dispirited, and disillusioned by the human race, my photo is of a crowd of people. People holding buckets of fish and coolers filled with raptor food, kids with plastic containers of bugs, friends and family with checks made out to Flyaway, Inc. Veterinarians in lab coats, fellow rehabbers holding carriers. Those who put themselves in harm's way to help an injured wild creature, the ones who drove hours out of their way to deliver it. The woman who brought me an injured egret and gave me ten dollars she couldn't afford to give. The boy who raked leaves so he could donate his earnings to help a dove.

Standing in the middle are John, Mac, and Skye, with ready grins and the gift of understanding.

And flying above them all, with his pirate's eyes and his single pale feather, is George. He brought me back, then he let me go.

ACKNOWLEDGMENTS

I wish to thank everyone who appeared in this book, all of whom have enriched my life in their own unique ways. I would also like to thank those who were a part of Flyaway, Inc., but whose stories remain untold, only because had I included every one, the result would have been too heavy to lift.

I wish to thank Ed Stokes for his long-distance friendship, and for his unwavering insistence that I write a book describing what wild birds are really like. Thanks to Sandy and Howard Hoffen and Karen Haskel, whose generous donations allowed me to avoid writing my yearly fund-raising newsletter and finish the second draft of the book instead. Thank you to Gail Winston, my editor at HarperCollins, for taking that enormous draft and shaping, smoothing, and whittling it into a better book, somehow finessing her way around my initial belief that cutting out any bird's story would desecrate its memory and drive a stake through my heart. Thanks to Sarah Whitman-Salkin and Shea O'Rourke, who patiently helped me through the details of publishing, to Leah Carlson-Stanisic for her elegant book design, and to Christine Van Bree for her lovely book jacket.

The beautiful illustrations by artist and fellow bird rehabilitator Laura Westlake were based, for the most part, on photographs of the actual birds who appeared in my stories. Although many artists can render a skillful depiction of a bird, Laura's drawings reflect her rehabber's understanding and appreciation of each bird's individual personality; for this, and for her friendship, I am very grateful.

ACKNOWLEDGMENTS

I am indebted to so many of my compatriots, but special thanks to Betty Conley, Elaine Friedman, Melissa Gillmer, Nancy Goldmark, Jennifer Gordon, Pat Isaacs, Ellen Kalish, Karen LeCain, Kim Lennon, Cathy Malok, Lynn Miller, Pat Nichols, Toni O'Neil, Vicky Pecord, Lia Pignatelli, Connie Sale, Meredith Sampson, Michele Segerberg, Giselle Smisko, and Marc and Diane Winn, all of whom have provided me, through the years, with help, advice, and species expertise. My gratitude to Marge Gibson, unsurpassed storyteller and founder of the Raptor Education Group, who somehow found time to read my initial draft and give me her invaluable feedback. I would also like to thank the following veterinarians, who have so generously donated their skill, time, and compassion to relieve the suffering of injured wild birds: Dr. Richard Joseph, Dr. Andrew Major, Dr. Anthony Pilny, Dr. Martin Randell, Dr. S. J. Schimelman, and Dr. John A. Wilson.

My heartfelt appreciation to Susan Landstreet, my partner in Ladies' Dinners and nonprofit stress management, to Chris Mineo, for his Thanksgiving Day rescue, to Chris Pellettiere, who fell heart and soul for a young redtail and renewed my faith in the public, to Sheila Shuford, my fairy godmother and a bona fide Wise Woman, to Jim Tyrie, who is always willing to harbor a runaway (and her crows), and to Joellen Wheeler, my friend and lifeline, who keeps me laughing no matter what.

I wish to express my ongoing gratitude to my agent, Russell Galen, who read my initial collection of short stories and gave me such a detailed critique— as well as such inspiration and encouragement—that I eventually produced a book I never would have thought I could write.

And my deepest thanks to my family, who fill me with love and wonder and make me believe I can fly.